Ha[illegible]

r[illegible]

from ka[illegible]

Nathalie.

THE DAYS OF THE GARRON

By the same author

REPRODUCTIVE BEHAVIOUR IN UNGULATES
Academic Press, London and New York, 1968

FARM ANIMAL BEHAVIOUR
Bailliere Tindall, London, 1974, 1975

THE DAYS OF THE GARRON

The Story of the Highland Pony

BY

A. F. FRASER

MACDONALD PUBLISHERS

ISBN 0 904265 26 9

Published by
Macdonald Publishers
Edgefield Road, Loanhead, Midlothian

Printed by
Macdonald Printers (Edinburgh) Limited
Edgefield Road, Loanhead, Midlothian

To

the memory of

my brother

NORMAN FRASER

CONTENTS

ILLUSTRATIONS

INTRODUCTION

My youth in Scotland was spent alternating between Glasgow and the Isle of Skye. At the age of eight, three years before World War II, I saw my first Highland ponies running out on the hillsides near Dunvegan, Skye, and was put on one to ride—a sturdy yellow gelding. The relationship was exciting but short-lived. Soon after, a fractured leg suffered in a fall from a dry-stone sheep pen required my removal to hospital in Glasgow. Four years later, living close to the blacksmith's "smiddy," I got to know by sight every horse in the parish of Strath. When the travelling stallion arrived from the Department of Agriculture in May to take up summer residence and breed the mares of the parish, it was like a visit by Pegasus to me. A grey horse each year, he glimmered with grooming; mane, tail, forelock and feathers flew as he danced in his red striped surcingle, his walker by his head, striding out in a fresh direction each day. The arch of crested neck, the sharp pluck of each foot, the tail carriage, the impatient "half-pass" gait, this was surely the ultimate in power, elegance and natural beauty. Would it be possible to work with the animal? Yes, I got to clean out his stall, and thus entered the service of the horse. My course in life was set. An uncle had a grey mare called Maggie—after his wife. The mare was believed to be the best-looking Highland pony in the south end of Skye. Maggie and I became somewhat taciturn partners in the work of the seasons—ploughing, harrowing, carting. At times she pulled the plough alone, sometimes she helped break in others. When we carted peat from a distant bog, she could put her shoulders into the collar and make a cart road where none had been before. Riding her in from common grazing without tack, using a hazel switch, was a thrill. She was boss in the herd of local ponies when they ran out in the winter on the common grazing. At times she was stabled or tethered on the croft. She would be crest-fallen and a little ill-humoured then.

Schooldays and wartime ended together, somewhat appropriately, and I entered veterinary college. Holidays were spent, whenever possible, on horse work. Sometimes in Skye, sometimes carting with Clydesdales in Lanarkshire. My first professional job was, naturally enough, in the Isle of Skye, but Maggie had gone now and with her almost all the horse population of the district. The blacksmith was retired and his smiddy was silent.

I moved south to agricultural work in the North of England but the Dales and Fell ponies were almost all gone, too, displaced by tractors. As years passed I could observe the gradual return of the horse, a sporting and recreational animal now. One day an old grey Highland mare came my way, as though she had chanced out of the past. I bought her and was back in active service with the horse. I learned of a Highland stallion for sale in Argyllshire, a young grey steed with familiar hallmarks. My return to the world of the Highland pony was inevitable after the stallion's purchase. Ponies were studied and visited throughout the Highlands and Islands. This breed had history. It led me to undertake a study in equine biology which became concerned with the histories of breeds and the relationships between horse and man over the years.

This research into the natural history of the horse in Caledonia has led me into many avenues of investigation. Many friends have been particularly helpful in drawing attention to historical works that are quite rare. Some of these have given interesting comment on the status of the horse at various periods in Scottish history between the 12th and 18th centuries. While these comments have not concerned themselves with the horse in any great depth, nevertheless the remarks on this animal relative to its participation in the activities of the Highland society and the historical evolutions of regions of the Highlands have been enlightening. Some of these documents have been minor publications such as the account of *The Highland Pony* by Thomas Dykes. Others have been interesting, indeed fascinating accounts of the close relationship between the Highland pony and Scottish social life such as given by MacDonald's account of *Highlandmen and Horses*. One particular historical item of literature which deals with Scottish history in very considerable detail has made very brief mention of the horse in Scottish affairs, but such passing references as have been made must be acknowledged since the historical work in question is probably one of the most definitive accounts of the history of the Highlands. This is a four-volume publication by James Browne, published in Glasgow by A. Fullarton in 1838. The extremely well-balanced and comprehensive Scottish history by Fitzroy MacLean and the classic work entitled *The Foals of Epona* by Dent and Goodall have served throughout to give a reliable historical backdrop to the story of the Caledonian horse.

The pursuit of this study has given me some justification for the travelling instinct which affects so many Scots. There have been horses to see in many countries. One winter's day, lecturing in a biology class in Canada, I was faced with a dilemma. The very heavy snowstorm overnight had depleted the student numbers. Should the scheduled lecture be given to less than a quarter of the class? They could not be turned away after dutifully attending; some had arrived on snowshoes! One student suggested a lecture on the ecology of the pony of the Western Isles of Scotland, a breed which he had encountered in his native Newfoundland. To the best of my ability, I complied. Pieces of the story began to fall into place. This book was started. Hopefully the story is now more completed and will serve at least as promotional material for this animal which, I believe, has something to tell mankind about living with nature, about surviving in harmony with the environment. This pony has certainly given much to Scottish history. It has a place, too, in Scotland's return to natural interests.

Chapter I

FEATURING THE HIGHLAND PONY

The Highlands and Islands of Scotland are rich in beauty and history. The land is geographically and geologically complex and varied; it is composed of mainland mass, islands and islets. These are honey-combed with lochs, glens, mountain ridges, shorelands and rivers. The country is stage propped with grasslands, crowns of rock, patches of bog, great carpets of heather and ornamental woodlands. Its weather is no less mixed. Rain can precipitate like champagne or a reminder of the Flood. Winds can be loving sighs of nature, or howling primeval ghosts. Atlantic currents carry West Indian warmth against Nordic chill in a changing but temperate balance. The climatic fortunes of a year are no more predictable than the weather of a day. In a constant battle, inclemency attacks and retreats from a shy, defensive sun. Local life is patient to all of this, content now in its biological niche, secure from past upheavals. Here is the source of the Gael's enigmatic taste for venture and his respectful appreciation of nature. Historically, the Gaelic races of man and animals have endured hardships. They have obliged the fortunes and misfortunes of time and circumstance.

The physical sciences have received intense attention from modern man, and repayment of this has been technical and material progress. But as by-products, there have also been monstrous environmental problems. The sundry life sciences have lately come into their own, almost as a counter-balance. The incomprehensible lack of harmony among mankind today is only a continuation of the human aggression which constitutes history. The story of Scotland provides a good example of this. The study of sciences such as agriculture, archaeology, history, geology, biology, sociology, ethnology, ecology and veterinary science can fit together. Sociobiology is their fusion, and knowledge from this source, with relevant faiths and beliefs, could give guidance to mankind for the conduct of our affairs. In keeping with its own faith in education, Scottish tradition tempts sociobiological studies on itself. The relationship between working animal and man in Scotland's past invites special attention. For example, survival clues may become evident when we examine closely the history and lifestyle of Caledonia's true and faithful servant, the Highland Pony.

Prehistorically domiciled with the Gael, in their joint homeland, was an indigenous pony whose ecology awaits exploration. The animal's merits of character deserve recognition. In partnership with a spartan people, this Caledonian horse has a history of adventure to fill the mind. It has, too, a future of popular recreation to be relished by those rediscovering nature and her ways.

Fossil remains dated 60,000 B.C. have revealed that a smaller predecessor of the Highland Pony flourished in Scotland then, to be lost in the last Ice Age. The pure-bred Highland Pony today has ingredients of various early horse types but its basic lineage traces to a primitive Northern race of horse. The changes of horse usage in the Highlands, brought about by depopulation of crofting communities, the introduction of engine power and the new recreation of nature trailing by pony, have resulted in changed values in conformation of the breed. Phases of cross-breeding, both by historical incident and legislative policy, have also effected minor breed changes. The types most numerous today feature the lighter "Western Isles" specimens and the larger, heavier "Mainland" type. "Garron" is the name, of Gaelic origin, for the gelding of either type. This name is erroneously applied, fondly and respectfully, to all Highland ponies, whether "cut" or not. It is a title more than a name.

The Highland Pony of today is the most versatile and powerful of all the numerous breeds of pony native to Britain. It is indigenous to the Hebrides and the mainland Highlands of Scotland. This hardy horse, of impressive endurance and kindly disposition, is of pony height (by definition, standing between 13 and 14·2 hands, or up to 145 cms). Grey and dun colours, of attractive shades, are most common. A dark "eel stripe" usually runs down the back, as a significant hallmark.

The Highlander should be screened by a long and thick forelock, which should be long enough to reach well down the face. The head is neat and comparatively small, with a well-shaped, deep and wide forehead. Its eyes are alert and expressive and overlook the concave, slightly dished face. The nose is straight and narrow and its narrow bridge improves frontal vision. Ears are hairy and neat,

Pony and boy resting on the moor

A group of garrons being tacked up: note the eel-striped backs

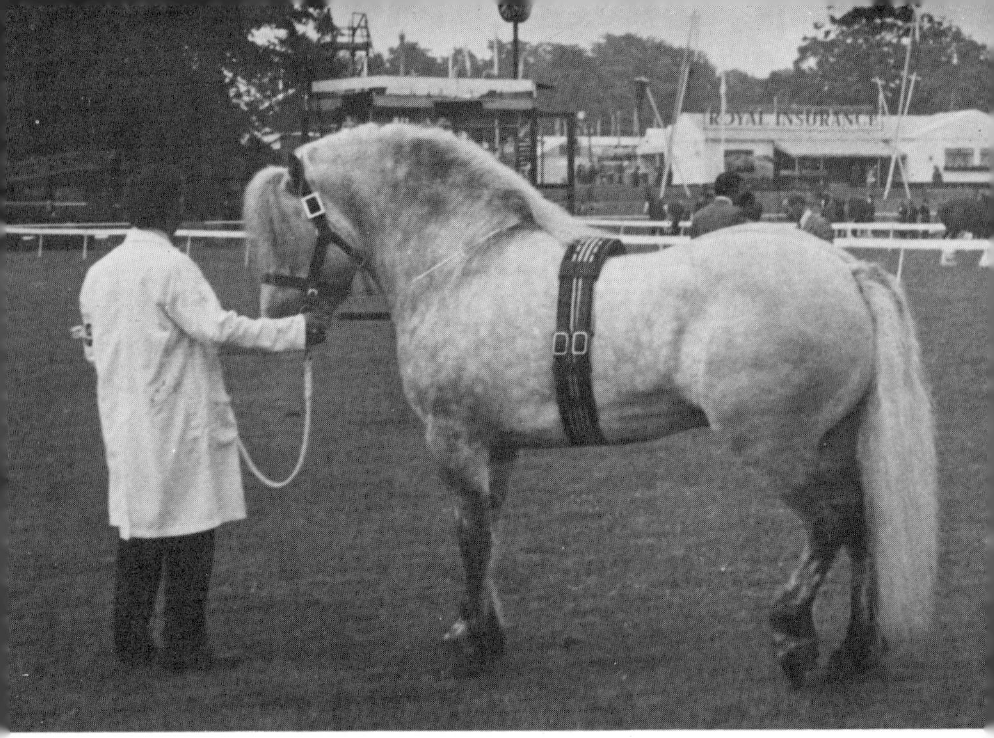

The compact power of a Highland stallion

Brisk gait in a Highland pony at the walk

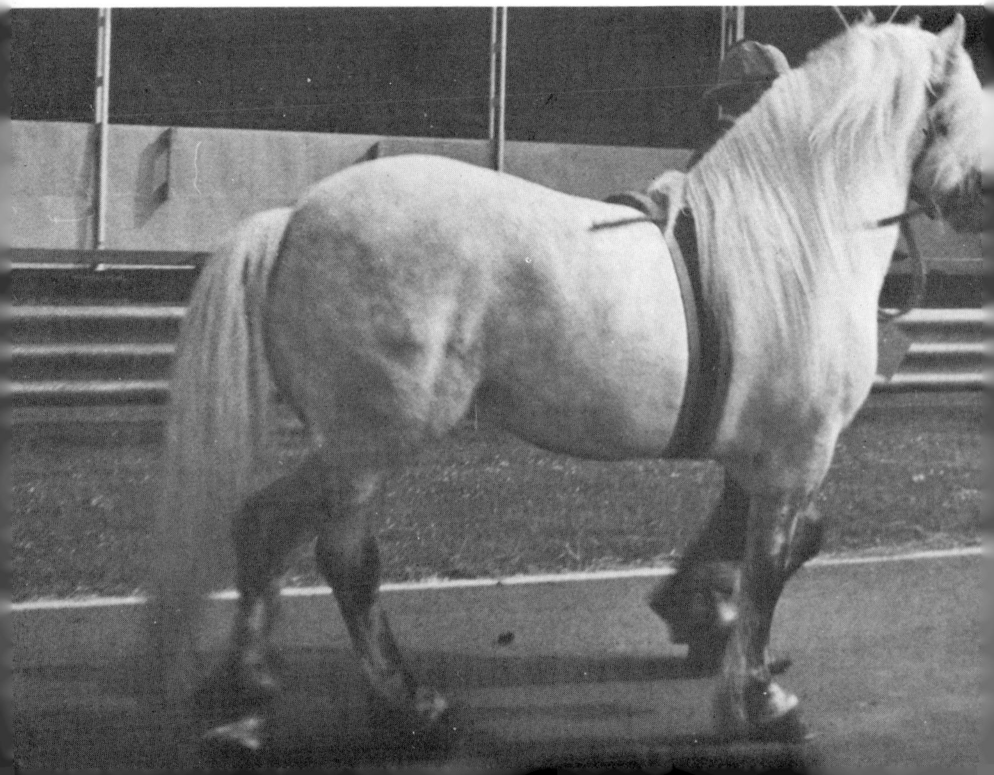

and the jaws are nicely rounded and strong. The muzzle is large and broad with wide nostrils. The neck is particularly well-developed and its crest is arched. Usually, the neck is well-developed at the base where it merges nicely with the withers, shoulders and breast. In good specimens, the back and loins are compact, and the tail well set-in.

In this breed, the chest is deep and well sprung. The forelegs are straight and from the front an imaginary line should pass from the shoulder joint straight down through the forearm, the centre of the knee, the length of the cannon bone, the fetlock and the toe of the hoof. A side view of the leg shows the elbows well closed-in and flesh-covered and the leg to the fetlock in the true perpendicular. The hind leg permits, when viewed from the rear, an imaginary vertical line to pass from the point of the buttock down to the point of the hock and the posterior of the cannon, in the perpendicular. From the rear, an imaginary vertical line passes from the buttock, through the hock and the middle of the fetlock, ending between the bulbs of the heel. The tail should be long and bushy, with wavy hair reaching the vicinity of the fetlocks and the tail root hair should have good body, especially between the buttocks.

Stallions usually have nine to ten inches of bone; mares have about eight inches. Short cannons and flat bones are good features. Good sloping pasterns, particularly in the forefeet, are desired. Hocks are very well let down and although hind quarters may be heavy, some are not. Hocks are neat. The feather over the coronets is light and bouncey and while covering the hoof head adequately, it should not carry far down the hoof. Hooves are rounded and open, more especially the forefeet. The hoof horn is dark, hard and of flinty appearance.

The overall appearance of the pony is one of well-coupled strength, nicely contoured. Beauty of build, of head and expression, is evident. The carriage of the head is upright and alert with large dark eyes. In movement, the horse reaches out well and lifts all its feet cleanly, both in the walk and trot. The attitude of the pony is reassuring, leaving a kindly impression of its disposition.

In Highland Ponies a golden or yellow dun colour was formerly very common. Colours now vary, but are solid. Grey is fairly common in a dappled shade, sometimes with a trace of the dark dorsal stripe. With other colours, such as light-coloured duns, mouse duns, creams and iron-greys, silver traces in the mane and tail are not uncommon. Virtually all duns have the dorsal stripe. Other colours include Palomino, black, brown and bay. Black limb points are quite common.

The dorsal stripe is almost invariably present in all pure-bred Highlanders, except in blacks, browns, iron-greys and some dark greys. The dun with the dorsal stripe is the primitive colour type. Grey may be considered mutant and has become a popularly selected colour in breeding sires. It is of interest to note that a grey horse named Sleipnir featured in Norse mythology as the mount of the chief god Odin. It was believed by Vikings to be equally capable of travelling over land or sea.

Colour has been mentioned in a way which suggests a fashion, but its selection may be based on a search for desirable traits, too. In some mammals, including horses, there is a rudimentary relationship between the pigmentation of the skin and certain hormones which can affect temperament. For example the pigmentation melanin and the hormone adrenalin share a common metabolism. They have the same biochemical precursor which is required for their production. A number of different experimental animal studies have recently shown that certain basic behavioural features such as fear and aggression can be influenced by breeding for a certain coat colour. It has long been firmly believed by a proportion of equine experts that there are some colour-temperament relationships in horses. Chestnut, for example, is a colour often associated with volatile temperament, while greys and blacks have greater temperamental stability. The dun colour, certainly in the Highland breed, is often associated with a phlegmatic disposition, and this may be the basis of the breed's reputation for good temperament. It is also significant that chestnut is a missing colour in this breed—conspicuous by its absence, in fact. The "suite of characteristics" of dun colour, eel stripe and sound temperament characterise the breed in a way which is not likely to be coincidental. Here, surely, is an outstanding example of a physical-temperamental relationship. Their deliberate selection is probably due, in large measure, to the wisdom of choice by those breeders who had to live at very close quarters, and in total harmony, with the working ponies they bred. No temperamental chestnut would find favour with them.

Another primitive and common feature is the presence of zebra markings, caused by bands of lighter and darker pigmentation. The result can be a patterned effect such as zebra markings on limbs, lattice markings about the shoulders or forearms, and general brindling. Lattice and zebra markings on upper and lower leg areas respectively can often be seen in old-fashioned, pure-bred Highlanders, and are evidence of primitive ancestry, surely a colour feature worthy of conservation.

Many of these features of conformation have become more important with the increased popularity of this animal in shows. Most, if not all, of these features of conformation are of functional value to the animal in its natural work and setting, thereby justifying their recognition in the show ring. Its conformation is a testament to its very long history and its adaptation over several millennia, to its Highland environment and Gaelic life.

Chapter II

A QUESTION OF LINEAGE

The study of the complex web of population genetics and physical features of animal life has become a very rewarding method for unravelling the twists and turns of joint human and animal history and prehistory. It also helps to solve problems of the origins and dispersal of animals. In each individual case study there are limitations. To obtain the most satisfactory findings and conclusions in such population studies it is therefore desirable to study socially-related subjects, or species, simultaneously. The close horse-and-man relationship has several advantages in this type of socio-biology. One advantage is that horses display very visible variability. The genetic bases for their variable traits are quite well understood. Most of this variability is in physical appearance, such as size, weight, conformation, coat colour, hair texture and gait. Some special features of behaviour, such as drinking habits and phlegmatic or volatile temperaments, also show distinct variability between general breed types and provide the common features of a general population. The accumulation of information on conformation and behaviour for populations is a relatively novel but inexpensive scientific procedure, though a time-consuming one. Although expensive technology is not needed, the findings drawn from adequately-stockpiled information can be assured of being sound, there being safety in numbers.

Ponies might not, at first glance, seem to offer great assistance in studying historical sociology but they can provide many clues. Horses and ponies use potentially arable land to which mankind is also attached. The pony has been spread from its early homelands to every inhabited part of the world. Isolated communities of ponies even continue to exist very successfully in areas abandoned by man. In addition to being ubiquitous, horses are numerous; there may be 100 million today. Sampling their populations is easily performed by studying reliable international equine literature and inspecting, through travel, good breed specimens on display. The equine biologist has a wealth of material with which to study the trail of the horse in history and in association with man. Adequate written records of history only began 200 years ago. Earliest writing, in any form, only goes back to 4000 B.C. Even that relates only to isolated populations in the Middle East. For many human societies, historical records of culture are comparatively recent. Any supportive evidence of prehistoric activities obtained from man's associated animals is very valuable circumstantial evidence of his early story.

The horse is senior to man as an established species. Pony life goes back ten million years, to man's two million. Pony history, both distant and recent, can be seen in a different light as an historical subject. The horse has had associations with mankind, perhaps from our beginnings. With the horse, man has had power to till land, to travel, to wage war and to symbolise economic and social trappings. The study of equine population is therefore rewarding, not only for what it reveals about its own adaptations, but also for what it suggests about any human population with which it is evidently associated.

After our primitive *Equus* first appeared, he became vulnerable to a predator in the form of man. Early cave man feasted on wild horses. The skeletal remains of about 100,000 horses have been discovered in a Cro-Magnon cave in France. This is evidence of the extent to which this man ate horse-flesh, slaughtering the animal for food on such a scale. Human intelligence eventually prevailed and man learned to put the horse to alternative, and better, use. Mounted on tame horses he became a warrior of a new dimension. The wheel allowed the development of the war chariot, the first offensive military vehicle.

The Hittites from Asia Minor were the first people to make full use of the horse in warfare. This allowed them to acquire the first empire. Persians followed suit, invading Greece. The Greeks then taught the Romans the effectiveness of the horse in warfare. The Roman Empire was finally destroyed by the Goths, mounted on the heavy horse, now indigenous to mid-Europe.

The horse was well-established through many areas of the world long before mankind had succeeded in doing so. The varieties of early forms of the horse have been found in archaeological discoveries, particularly throughout Central and Northern Europe. Through these findings the evolution of the horse can be traced.

It is clear that various forms of primitive horse were associated wtih specific regions of the broad European continent.

To the east of the continent, in Mongolia, a primitive horse existed. That horse is now named the Przewalski horse, an equine species which has been successfully preserved after reaching the brink of extinction. The central stud registry of this virtually prehistoric horse is kept in Prague but specimens of the species, numbering a total of 272 in 1977, are maintained in zoos over the world. One pair of stallions are now in Kingussie.

In Central Europe the primitive horse form was the Tarpan. Although this horse seemingly became extinct several decades ago, Polish scientists have succeeded in capturing wild horses in that area of Europe which evidently contained strong traits of the Tarpan. By careful selective breeding, they managed to re-establish a horse which is essentially Tarpan. It is preserved now as a "protected species." Both the Przewalski and the Tarpan are dun to brown in colour and have a black dorsal stripe! They are pony sized. It is considered that they were the early forms of the warm-blooded horse, strains of which include the modern Thoroughbred and its close cousin, the Arabian. The latter, of course, is the elder breed and is the blend of many types which had their earlier origin in Mediterranean countries.

Descendants of the other principal form of the primitive horse can be found through the remainder of the horse world today. These are the cold-blooded horses, which have two forms. Most typical are the heavy draught horses which appear to have evolved in Western Europe, particularly in France and the Low Countries. The other smaller form is characterised by the Nordic horse which has a variety of types, each having certain similar features. These are found throughout many countries of Northern Europe from Iceland, Scotland and Norway through Sweden, Finland and into Russia.

There has been much speculation about the origin of the Scottish Highland pony—attempts to establish its historical identity have been based on examination of some archaeological findings and certain specific characteristics. What no-one appears to have done is to have made a serious examination of any possible close relatives of this breed, based on comprehensive similarities and more general characteristics and to study these wherever they occur elsewhere on the globe; to trace, as it were, the hoofprints of this breed's unrecorded ancestors.

When this type of lineage search is made, a clearer picture emerges of the way in which the Highland Pony's history apparently progressed. It is fairly certain that the pony was established as a primitive breed in Scotland before the country was properly inhabited by mankind. This horse, however, was not the same as the modern Highland Pony, which is a mosaic of several early types and blood lines which have come together. Thus the component foundations of the Highland Pony come from several ancestral sources, giving it mixed ancestry. The basic ingredient, however, of this animal is evidently Nordic, still very pre-potent, and providing it with its principal characteristics.

By tracing Nordic strains of horse breeds across Europe we find noticeable similarities in colour, morphology, in temperament and behaviour. As we search backwards against the accepted route of equine migration from central Siberia, examination of the Nordic type of horse leads to the horse in Poland. Initially, however, such an examination centres on Scandinavia, where one can pick out a Nordic horse in the form of the Northland breed. According to tradition, the ancestors of the present Northland horse were introduced from Russia over 1000 years ago. Very little original information exists about this early horse but it is clear that it was a small animal, sufficiently small to be called a pony. It was used for riding and for draught work in northern Scandinavia for hundreds of years. The breed was used mostly by farmers who made no attempt to standardise or modify the breed until very recent times.

A neighbouring breed of horse is the North Swedish horse. This breed originates from an ancient one native to Scandinavia and undoubtedly related to other localised breeds, particularly the Døle or Gudbrandsdale horse. The North Swedish is also very closely related to certain Norwegian horses. Since the beginning of this present century, the breed has been improved by an official breeding programme in which only licensed stallions are used. Breeding stock must combine strength and liveliness of gait. Another characteristic of the breed is its excellent temperament and its great courage in facing heavy draught work. Another close relation of these horses, having its place in the family tree of the Nordic strain, is the Fjord pony of Norway. It has been an inhabitant of

Norway since prehistoric times. Vikings are known to have kept and bred this type of pony, and there are prehistoric carvings in stones which show essentially similar forms of horses over a wide area of Norway. The smaller Fjord pony is now the most popular and is well suited for the work on its own terrain where the going is difficult. The Fjord pony is small and stocky, has a thick-set neck with a well-set head, small pricked-up ears, short back, strong hind quarters and a deep chest. It is fairly short of limb and carries a slight amount of feather on its fetlocks, and a considerable length of tail. A most common colour is dun with a dorsal stripe.

The Gudbrandsdale is a northern Scandinavian breed of horse native to the eastern coast of Norway, although distributed throughout the whole country. It is very similar to Swedish and Finnish native ponies in physical capacity and conformation generally, but variable in size, often being larger than pony-size. They are usually dun in colour with a black dorsal stripe. They carry long manes, tails and fetlock feathers. Hardiness is a feature of the breed. It lacks the good hindquarters of the Highlander but could be passed off as one in a crowd.

The Døle horse of Norway is a very handsome breed. It shows many of the characteristics of the Highland Pony but also some conspicuous differences. Probably it more resembles the British Dales breed of pony. Nevertheless, there are features in this breed reminiscent of the garron, features such as the size, colour and elegant shape of head. A common colour is bay.

Another prominent member of this equine line is the Finnish horse, bred from ponies native to that country for innumerable centuries. These ponies are very similar to others still to be seen in other parts of Scandinavia where various types and sizes exist.

The most common type is used for riding and light transport and for agricultural draught work, being noted for its toughness and working capacity. A common colour is mouse-dun, with a silver mane and tail also being a colour feature.

In Poland there is the Konik breed. This horse is usually about 13 hands high and is described as being more of a horse than a pony. The Konik is known to be able to thrive and work hard on a meagre diet. This makes it popular with the small farmers in Poland and throughout the adjacent parts of Eastern Europe. It is also noted for its robust constitution and its very amiable and quiet temperament. When the Konik horse is studied on the hoof some similarities to features of the Highland Pony are inescapable. It has the compactness of build, the set of the neck and head, also possessed of the Highlander. It carries a long, thick mane and tail with silver streaks throughout these; a common colour is mouse dun. The legs are usually dark.

The Yakut, from a first-hand report, is a most remarkable breed of horse. It has its origin in the valley of the river Yana in the central part of northern Russia, but its territory extends widely and includes the polar circle to the north, containing some of the coldest regions of the northern hemisphere. Also included in the homeland are grasslands to the south which have deep snowfalls in winter and short, hot summers. The Yakut horse is an excellent pack animal. It is also used by local people in recreation, is a competent racer and is also used in a variety of equestrian games. The horse varies in size and is approximately of the same height and weight as the Highland Pony. The usual colours of the Yakut horse are reported to be light-grey, greyish or mouse-coloured. Some have dark, latticed patterns on their shoulders and the tell-tale eel stripe can be found. These, of course, are the classical colours of the Highland Pony and of Nordic horse kind in general. It could be strongly suspected that this breed represents the origin of the Nordic strain.

Traces of this kind of pony can be found from the Yana valley westwards but in some parts of Europe, such as Western Russia, which has been the location of so many wars, this horse is reported to have vanished. Earlier forms of warfare took a colossal toll of horses. Following wartime it was usual practice not to re-establish the horse population using the same native breeds but to attempt to establish newer breeds locally by introducing foreign blood stock. These were often selected from quite distant countries in order to establish more distinctive breeds of horses of more desired characteristics. Such policies removed relics of the indigenous horse types of that expansive region of northern central Europe.

Nearer to home, the Icelandic pony is a close cousin to the Highlander but has a much less complicated history. It is generally believed that the ancestors of these ponies were not indigenous to Iceland, having been introduced by the original settlers from Norway in the ninth century A.D.

A point to be borne in mind is that there must have been importations of the Scottish Highland Pony to Iceland, so that the Icelandic pony, in resembling the Highlander, has this good reason to do so. The Icelandic pony is generally smaller than the Highlander, standing between 12 and 13 hands. It is usually used for riding and has more of an amble in its gait than is typical of the Highlander. These short and stocky animals are noted for their good disposition and hardiness in living outdoors in severe weather.

The late Professor Ewart of Edinburgh, in studying the Icelandic pony and some of its obvious close relations, was tempted to speculate that there must have been a prototype of this form of horse which he called *Equus Celticus*. He claimed this to have been the common ancestor of all the pony breeds of northern Europe. There would not be too much to argue with if that claim specified north-western Europe excluding Scandinavia. But it does seem that the claim may have been too expansive. As an early isolated prototype, its contribution to the Highlander would have to be balanced with blood from later Roman horses together with the clear input of ancestral blood from Scandinavia and from Spain. It might be reasonable to acknowledge the existence of a Celtic pony of an earlier primitive type, but specimens of horses which can be examined today in various northern countries clearly indicate that the Nordic was the basic type of primitive horse to be found there. Doubtless, it was well-appreciated by many of the peoples who subsequently came to inhabit these various countries. As mentioned, in Norse mythology the chief god Odin had a grey horse called Sleipnir, a steed which could travel over land or sea in Viking lore. Perhaps the sight of a familiar horse on their Scottish excursions encouraged the Vikings in this belief.

The careful sculpture of prehistoric carvings found in Norway and in Scotland clearly indicate the reverence these early peoples had for the horse. Indeed, it comes as no great surprise to learn that the earliest indication of the use of the side saddle for women appears on a carving on a Pictish stone in northern Scotland, dated about 800 A.D. The carving is a monument showing a Pictish lady of culture, possibly a princess, riding side saddle on a pony type which can quite easily be recognised as an ancient ancestor of a garron.

CHAPTER III

ENTER THE GARRON

The environmental characteristics of the Highlands and Islands of Scotland give the region its own special features created by earthquakes and volcanoes. Ice Ages carved mountains and deep lochs of sea and fresh water, forming glens and massive areas of stone rubble. The dramatic sculpting of the land came to an end with the passing of the last Ice Age when all this part of Scotland settled to the general form it has today. During the third period of the Ice Age, however, Scotland was joined to Scandinavia by an ice cap.

There is no written history, of course, which gives any account of these changes, but geological and archaeological discoveries supply sufficient evidence for us to understand in general how these Highlands and Islands took their shape.

Preserved and embedded within rocks and fossils is the evidence that Scotland is a very ancient habitation. Evidently cave dwellers in the early periods of habitation were people who hunted and fished; stoic people who survived amidst the hardships, the climate and the hunger.

Later, Scotland welcomed a primitive, neolithic people. These were adventurers travelling by sea from Mediterranean areas in the south, from the continent of Europe across the North Sea, and from Scandinavia. Relics of ancient deep-sea-going boats have been found in the estuaries of Scottish rivers; boats which date prehistorically. These immigrant people brought the Bronze Age to the Scottish Highlands. About this time Britain's location on the globe may have altered slightly. This geographical shift, however slight, appears to have been sufficient to modify the Scottish climate, making it more temperate, bringing great stretches of forest to the northern Highlands and some Islands and encouraging the lush growth of summer grass in meadows and glens. The climate became stable and the land more habitable. Early migrant man settled now in great numbers in this land which came to be known as Scotland.

It cannot be stated accurately who the initial immigrants were. The only certainty is that they were of diverse origin from Western

Europe and Scandinavia. The earliest reliable account of the nature of early immigrants to Scotland is the migration of the Celtic civilisation which moved from central Europe westwards towards the Atlantic coast, into Brittany and Spain; across the sea to the Cornwall peninsula; northwards through Wales and Ireland and finally settling throughout the broad expanse of the Highlands and western coastline of Scotland, the outer limit of Europe. This Celtic migration extended over a period of time and carried with it its own Celtic civilisation as it spread westwards. Its active migration is estimated to have occurred in the sixth century B.C. Within 300 years the force of the Celtic migration had reached Scottish shores, and from that point of history onwards this Celtic race and the remnants of their culture have formed the basic fabric of Scotland's population. Characteristics commonly observed among Scottish people are still attributed to the Celtic race.

The early Celts were described as high-spirited, frank, with a keen sense of honour but traits of barbarous savagery and much given to religion, philosophy and poetry. In addition, according to the writings of Caesar, they were warriors who used horses with skill.

The culture of this people had an artistic element. Innumerable relics have been found of Celtic jewellery in the form of fine silver brooches, gold pins, jewelled rings—many of these with elaborate ornamentation. Further archaeological findings include artifacts such as decorated pots, weaving combs, wooden writing tablets, metal styluses as writing instruments, leather tents, tools for carpentry and agriculture. Significantly, there were items for horse-shoeing, harnessing, and wooden carriage wheels. Clearly these people were not only a cultured and organised race but they were an agrarian people who used horses and horse power. This was the nature of the indigenous people of Caledonia, the Scottish Highlands and Islands, when the Romans arrived in Britain.

In 80 A.D., 37 years after their conquest of England, the Roman legions first advanced into Scotland. The people of Caledonia arrested their progress and held them south of the Highland Line. This is the phrase which now defines that major geological fault north of Stirling, running diagonally north-east to south-west, making a mountain range with several distinguished Bens. There has always been a Highland Line of a metaphorical nature above the waist of Scotland, for the Highlands and Islands have always maintained some form of social boundary between themselves and the remainder of mainland Britain. About this time, Scotland's population became recognisable in two main groups, the Picts and Scots. In partnership they held the Roman legions at bay. Of these two groups, the Picts were probably the ones most representative of the earlier aboriginal inhabitants of northern Scotland, the very early Bronze Age colonists. Much of Scotland's history in the post-Roman era concerns the negotiations between them to live in harmony and to form one common nation.

In Browne's *History of the Highlands and The Highland Clans*, Vol. I, he describes that when Agricola invaded north Britain in the year 81 A.D., it appears to have been a land occupied by 21 tribes. These groups had very little political connections with each other although they were evidently people of the same nation, with a common language and social customs. He notes that the topographical position of these "Caledonian tribes" or "clans" at the epoch in question could be recognised and identified, that the country at that time was without agriculture, and was studded with bogs and woods, almost in the state in which it had been formed by nature. The Gaelic language was universal and was the common tongue among the people "on the Highland side of the boundary." This geographical generalisation has been taken, by this writer at least, as being the part of north Britain which can most truthfully be described as Caledonia as it then existed. Browne himself adds that Caledonia inhabited the whole of the interior of the country from the ridge of mountains which separates Inverness and Perth in the south to the hills which form the forest of Balnagowan and Ross in the north. This territory formed a considerable part of the extensive forest which in the early ages spread over the interior and western parts of the country on the northern side of the Forth and Clyde and to which the British colonists, according to Chambers, gave the descriptive appellation Celyddon, signifying literally the coverts, and generally denoting a woody region. It was on this account that the large tribe in question were called Celyddoni, a name afterwards Latinised into a more classical appellation of *Caledonii.* The name Celyddon, restricted originally to the territory described, was later extended to the whole country on the northern side of the Forth and Clyde, and this final general view of the range of area collectively termed Caledonii is the one which has been adopted throughout this book. It will be apparent that the term embraces the Western Isles as well as the mainland north of Loch Ness.

Caesar himself gave a description of Caledonians as a remarkably hardy race, capable of enduring fatigue, cold and hunger. They were decidedly a war-like people addicted, like the Hittites of more ancient ancient times, to robbery. The description continues: "The weapons of their warfare consisted of small spears, long broad swords and hand daggers, and they defended their bodies in combat by a small target or shield." Caesar's statement then includes the following: "The use of cavalry appears not to have been so well understood among the Caledonians as among the more southern tribes; but in battle they often made use of cars or chariots which were drawn by horses of a small, swift and spirited description."

Some historical records of Scotland's settlement were established by Julius Agricola, the Roman governor in north Britain in the year 81 A.D. These documents make it clear that Agricola in his campaigns into Caledonii made considerable use of cavalry and in the variations in military fortune he was obliged in many cases to retreat from furious attack in rather less than good order. In such circumstances, loss of cavalry undoubtedly took place to the Caledonian peoples. One of the strong stations established by Agricola to set up a defence against marauding forces from the north was in the higher part of the village of Carnock, near Dunfermline in Fife. At this site in Carnock archaeological findings have made it plain that this was a fortress of some consequence. The author, having lived in this region for a period of four years and having sought in the area for relics of Roman activities of that time, encountered small portions of weaponry and one complete horseshoe of a type which would be in use by Roman cavalry at that time. The size of the shoe would indicate that in all probability it would be from a horse of approximately fourteen hands high, although this is a rough judgment since the shoe was very poorly shaped in comparison with the standards of farriery in later centuries.

Caledonian peoples were offered Roman citizenship and whereas many may not have availed themselves of the opportunity to do so, it is distinctly possible that many learned certain Roman styles of living and some Roman skills including equestrian skills during eras of peaceful co-existence between the Romans and the Caledonians. Periods of peaceful co-existence were short lived, for the Caledonians avoided coming to a general agreement with the Roman emperors and kept up harassing warfare against the occupation forces.

It was clear that the people north of the Roman defences were involved in agriculture. Like farmers the world over, they had an independent spirit; it could not be broken by Roman aggression. Indeed, they turned from a defensive to an offensive attitude when faced with the numerous attempts by the Romans to occupy their country. As a result, Hadrian's Wall was built by the Romans across Britain to prevent the Caledonians of the northern part of the country from invading the Roman provinces. Later Antonine's Wall was built from the Forth to the Clyde, again erected defensively by Romans as another Scottish limit. More than one Roman Emperor attempted to extend the boundaries of his Empire to include all of Scotland. None succeeded. Some died of the exertions. More than one legion of experienced Roman soldiers became swallowed up in vain attempts to occupy the Northern part of Scotland. Romans used cavalry in these invasive attempts. With their defeats, many of their horses would undoubtedly be left in the field to be captured, to be used and mixed with the indigenous horses.

Although Scotland was never conquered by the might of the Romans, never experienced their habits, culture or civilisation, Scotland almost certainly acquired their horses of Mediterranean origin. If the horses resembled their statues, still to be seen abundantly in Rome today, this must be regarded as a great equine legacy.

After the Roman confrontations, the era which followed was by no means one of complete national harmony in Scotland, and subsequent centuries saw times of great violence and confusion over changing boundaries. Finally in 843 A.D. the Picts and Scots united under one King, Kenneth MacAlpine. He inherited a country well able to defend itself, peopled with competent agrarians and soldiers, who used the horse in the fields and in battle. In the background throughout this time, it is clear that the horse of the Scots was of critical importance in the development of the country.

Where had this horse come from? What was its type? The carvings of horses of these early ages in Scottish history can be found on silver brooches and on stones. It requires no imagination to see that these animals were the distant ancestors of the present day Highland pony, similar in size, *i.e.* somewhere between 13 and 14 hands. The horse was apparently strong and well-built in its hind quarters and fore quarters, a horse with a strong and arched neck and a long mane.

The early immigrants to Scotland, arriving by boat, assuredly brought horses with them. They would not intend to attempt to settle in this foreign land, with its difficult terrain, without ensuring themselves of a reliable beast of transport. Early Scottish horses, therefore, almost certainly included some from the Scandinavian countries brought by Vikings who were probably the most frequent visitors to Scotland's shores. These migrant adventurers were certainly very competent travellers who would find no difficulty in transporting horses with them for their periodic excursions to the Scottish North with the typical pattern of Viking visitations, and logically would be left behind. Having come from an essentially similar type of habitat the horses would find very little difficulty in adapting to the geography and climate typical of the Highlands and Islands of Scotland. They would also find appreciative people to tend them and to utilise them.

The Nordic horses as individuals would be added to the main equine population of Scotland at that time, which was predominantly *Equus Celticus*, as discovered by Professor Ewart and mentioned in the previous chapter. This horse had become established as the most indigenous to Scotland, having had various components in its family tree.

The Celtic migration itself had brought to Scottish shores horses which had been carried by the great trek through Europe, ultimately to the northern and western limits of the continent, notably the Highlands and Hebrides. *Equus Celticus* was apparently the fusion of the migrant horse population with the indigenous native horse of pre-history.

Little is known, unfortunately, about *Equus Celticus*, although we are grateful for Professor Ewart's work, establishing that such a horse not only existed but that it was the principal form of horse in Scotland in the Celtic era. It would be reasonable to assume that it was a hardy pony of an undifferentiated type, and one obviously suited to the agrarian culture of the Celtic people, who

Photo by courtesy of Mrs Margaret Fay Shaw-Campbell of Canna

A Barra pony and company: the resemblance to the Fjord pony is remarkable

Pictish warrior on horseback

Pictish stone carving of an equestrian scene showing side-saddle riding

Stone carving of Roman soldier on horseback
in Caledonia

Bruce and de Bohun: the encounter
central to Scottish history

A very fine type of garron on the Isle of Mull, a source of much of the best blood of this breed for a century

Highland pony muscled for draught work

required horses for the great variety of roles involved in their travels, periods of residence and agricultural activities. In other words, it would have needed to be a general purpose pony, doubtless stocky, strong and of a disposition allowing it to be a close companion to man. These characteristics would almost certainly be emphasised by breeding selection which would be quite ruthless in culling unsuitable individuals.

These two early types of horse, Nordic and Celtic, would be similar in size. The Nordic strain horses might have been a more muscled animal but they were of similar type and they had one critically important feature in common, that of being highly domesticated, an invaluable asset at this early stage of Scotland's settlement.

Ninth-century carved stones in the Northern Highlands showed the inhabitants as competent horsemen mounted on well-built, but medium-sized horses. They also possessed very sophisticated and decorative armoured headshields for their horses in the form of very finely beaten bronze. Some of these items are variously dated between 200 B.C. and 800 A.D. The stone illustrations and the nature of the armoury indicate these horses probably drew the war chariots of the Celtic chieftains.

Celtic culture is shown by archaeological findings to have followed progressive pastoral development. The communities settled permanently in localities of their own. Within these regions they used local resources, evidently grazing livestock on available grasslands. They established permanent buildings, using stone, and erected stone monuments of uncertain significance, but evidently relating to symbols of their emerging culture. A form of rural civilisation was in fact being pioneered in these settlements studded throughout the country. These small independent communities, did not use slave-labour, and in their ambitious stone-shifting enterprises would find the answer in horse-power. All these progressive developments in which they advanced their culture took place over several centuries, perhaps a millennium.

The Norman conquest was to have an impact on Scotland similar to that in England. Norman aristocracy mixed with Celtic aristocracy. Friezes and woodcuts of the era indicated that the indigenous horse was pressed into sundry uses in peace and war. One particular equestrian incident proved pivotal in the course of the country's history.

Robert the Bruce, of Norman ancestry, and of great equestrian skill, fought horseback, evidently on an earlier type of Highland pony, in a skirmish prior to the Battle of Bannockburn in 1314. The success of this equestrian contest to the death with the adventurous Sir Henry de Bohun—the latter fully armoured and mounted on a heavy warhorse, with Bruce on his smaller, courageous pony—undoubtedly inspired the Scottish force. The signal victory of Bannockburn, to this day, is the corner stone in Scotland's basic unity. From that time on, patriotism in Scotland has never died.

At the time when Scotland's independence as a sovereign nation was established after Bannockburn, the country probably possessed a large population of general-purpose horses, suitable for riding, drawing vehicles, and working on the land, horses that would be small enough to keep economically and to be worked by all the members of the family engaged in the small farming units, some of which composed crofting communities. The Highland Pony had emerged, but further refinements were still to be added to this animal by subsequent events.

One of the most dramatic passages of European history is the Spanish Armada, the culmination of long-standing friction between Spain and England. Queen Elizabeth of England pursued an aggressive policy against King Philip of Spain. In the early 1580s Philip of Spain started to plan the invasion of England with an army to be transported by the Armada, a huge fleet of ships intended to invade and occupy England.

The Armada was completed in 1588 and departed from the port of Lisbon in Portugal which was then under Spanish control. Over 150 ships comprised the Armada—many of these were large military ships carrying military supplies, men and armoury; they were not typical men-o'-war, certainly not battleships. It was therefore ill-assembled for a major sea battle which was not its prime intention, its plan being one of invasion.

An Armada ship would carry only top quality horses, including officers' chargers. It has to be remembered that the Spanish Armada was both a fleet and an army—an army that was supported with cavalry. This cavalry would be well mounted. The equestrian skills in Spain at that time were of a very high order. Mounted troops used Barbary and Andalusian horses of great quality. These and similar horses were abundant in Spain; they were the legacy of a much earlier invasion which Spain itself had experienced when the Saracens arrived bringing with them large numbers of Barb and Arabian horses. The influence of these two breeds of horses was great on the native Spanish horse, which also derived blood from Arab and Barb strains of horse. The Kladruber and Andalusian breeds would have been the basic type of horse selected by Spanish mounted troops. This Iberian horse was light, clever and sure-footed. Sometimes they were crossed with the Barb. It was this backcrossing that defined the Andalusian breed eventually, the breed taking its name from a province in southern Spain.

The Andalusian is a strong-bodied horse, normally having a good forequarters, with a deep body, well-muscled and high hind-quarters. It has a short back, good bone with a handsome head, well-defined neck with a good crest and it has an excellent temperament. A common colour is grey. The Spanish Armada was almost certainly loaded with this type of horse.

The Spanish ships encountered the English warships in the Channel and an engagement took place between the two naval forces, about 60 warships from each side participating. The English squadrons won the engagement, sank two Spanish warships, and damaged all of the remaining 58. The Armada was defenceless and crippled. It fled into the North Sea, seeking its way homeward by navigating around the north of Scotland and down its west coast. On August 14th, an order went out to the battered Armada fleet to throw all horses overboard since the water rations were depleted. A few days later British Admiralty Intelligence had received a report of a sighting of hundreds of dead and drowning horses and mules in the waters at a latitude of Aberdeen. Evidently, most of the Spanish ships had obeyed the order. It is clear, however, that it was either not received by all ships or was ignored by some. Indeed it is possible that one or two prized animals may have been deliberately reprieved and kept in defiance of the order.

When the Armada entered the Minch and the waters of the west coast of Scotland it ran into tempest. The Minch, of course, is a notorious area of water, much respected, and indeed feared, by the sea-going peoples of that region who are familiar with its dark and changeable moods. The crippled Armada was by now ill-equipped to deal with such storms and some of them sought shelter close to the numerous islands composing the Hebrides. In this run for shelter some were wrecked on the foreshores and rocklands beyond the coastlines. No doubt others gained refuge in harbours, never to complete the voyage. Other ships struggled through these storms only to founder off the Irish coast. Only 67 ships of the Armada returned to Spain and it can be readily calculated, therefore, that 61 Spanish ships were lost in the waters off the Western Isles of Scotland and in the Irish Sea.

One Spanish galleon sought refuge in the harbour of Tobermory Bay on the Isle of Mull, where it anchored for many days, before it sank from an explosion said to have been caused by a time-bomb planted in her magazine by an agent of the English Secret Service. The galleon apparently stayed in Tobermory Bay while the Captain carried out dealings ashore for provisions and water. A local legend exists in the Isle of Mull in which a stallion is reported to have escaped from the galleon to swim ashore the few hundred feet from the anchorage. The stallion is said to have been rescued and used for breeding. This legend has existed in Highland lore for centuries.

In 1955 an expedition made an organised attempt to locate the galleon in Tobermory Bay. The ship was eventually found in pieces, buried in silt. Divers raised portions of its lead-armoured hull, cannon-balls, daggers and scabbards. Parts of an equine skeleton were also raised from the wreck. Newspapers of that date carried full reports of these findings, which were eye-witnessed by many. This is evidence that some Spanish horses did survive up to this point in the Armada's passage, and there is no reason to doubt now that the co-existing legend is not correct. Even one stallion of Andalusian breeding would have been sufficient, if given an opportunity, to have had a profound effect on the genetic properties of the Highland Pony of the day. There remains the possibility that other horses also succeeded in getting ashore from other ships in the Western Isles, although this matter will remain a contentious issue.

Between the islands of Scotland the traffic, including horses, has always been quite active. Very few of the numerous Hebridean islands have lived in a state of remoteness in relation to neighbouring islands. It was the custom to exchange breeds of livestock including horses between islands from time to time. Wherever Spanish horses might have been beached, their type would be later dispersed through the Hebrides in active trading activities.

By the end of the 16th century, therefore, the Highland Pony had acquired all its major genetic ingredients that were to be added to its fundamental primitive blood. Subsequently, the Highland Pony has been marginally affected by the introduction of other types of horses, as enterprises by private landowners seeking to introduce some novelty into the breed, but the basic characteristics of the breed were now established.

During later centuries very significant hostilities took place throughout the length and breadth of the Highlands of Scotland. For example, it is reported that in a confrontation between the Earl of Argyll and the Earl of Erroll, the latter quickly engaged a select body of about 100 horsemen, being gentlemen "on whose courage and fidelity he could rely." The forces grew after this point and it is said that nearly 1500 men, almost all mounted, were mustered under the rebel Earls of Huntly, Erroll and Angus to oppose the advance of Argyll. Huntly received intelligence that Argyll was on the eve of descending from the mountains to the lowlands which led him the following day to send a party of horsemen to reconnoitre the enemy. Their findings encouraged Huntly and his men to attack Argyll after he had passed Glenlivet. Argyll was taken by surprise and, apprehensive that his numerical superiority and foot soldiers would be no match for Huntly's cavalry, he disengaged. He then awaited greater cavalry forces to join him, realising in his wisdom that horse forces in significant number were an essential part of a victorious army in the Highlands.

This military incident, only one of very many hundred military aggressions which characterise Scottish history, took place in May 1594. Huntly's smaller army was thrown into the battle, their hopes mainly dependent on the power of their cavalry which charged the enemy, causing wild confusion and flight. The lesson was there to be learned, but it is doubtful if it was driven home. During the remainder of Scottish history it is not evident that particular emphasis was placed on the force of cavalry to support the fierce aggression of the Highland foot soldiers.

This particular incident, apart from the evidence it provides of the value of cavalry in warfare, was also an indication of the quantity of horse power which was available throughout most areas of Scotland both for warfare and agriculture.

CHAPTER IV

THE ADAPTATIONS

The primitive ancestor of the horse was the Dawn horse. Much has been written about this animal, which has been found in fossils throughout Europe, and about the subsequent forms of horse that evolved. It is not intended to go into these evolutionary changes to the horse in dealing with environmental adaptation. The adaptation which will be considered here is that which the modern form of horse has made to its general environment, including the climate in which it customarily finds itself.

One of the most important considerations in adaptation is vision. The eyes of the horse are arranged so that it is able to see almost entirely to its rear and completely on both sides all at one time. This spectacular panoramic or scanning capacity of the horse's vision was undoubtedly a key feature in its early adaptation. The range of early forms of predators with which it obviously coped, was considerable. This evolutionary type of adaptation is not without some disadvantage under modern circumstances. Some of the unfortunate consequences of this vital system are the horse's inability to see exactly what he is eating. He is unable to focus his eyes close enough to see objects very keenly if less than four feet from his face. The horse has particular difficulty with visual "accommodation," that is, in focussing objects directly in front of him. When he concentrates to focus his eyes forward to his maximum ability he appears momentarily to lose the ability to observe consciously to the rear and to the sides.

The advantages of rear vision, in circumstances where the horse was subject to predation, are obvious. Most breeds of horse have their eyes so laterally placed that their rear vision is only blocked by the width of the horse's own body. This vision, together with his keen awareness of sudden movement and his preparedness for quick flight, made the horse, when he was a hunted animal, difficult to apprehend by potential predators.

These peculiar advantages and disadvantages of the horse's vision explain a good deal of its behaviour. Normally, the horse has monocular vision, that is, it has the ability to see separate things with each eye at the same time. This is the result of the horse having his eyes set on the sides of his head. As a result of this he may be able to see an object with one eye only for a while till adjustment of view brings the object into the view of the other eye also.

One significant effect of this is that minor objects, as they come into view, appear to jump suddenly into the horse's conscious vision as the second eye catches up with the first eye to observe properly a given object. It is this visual characteristic which explains the horse's inclination to shy from minor objects. This also explains why they are more likely to shy from moving minor objects less than they do from stationary ones; moving ones sail into view, while stationary ones jump into it and into sudden consciousness. Alteration of head position may also be required to bring a stationary item into view with inevitable sudden awareness. Again, the smaller the object the more the horse is likely to have this object come unexpectedly into his vision and cause him to shy.

Defects in forward vision are compensated by flexibility of the neck. The horse actually uses movements of his head and neck to help focus his eyes by raising or tilting his head to one side or the other. The retina of the eye, the part of the interior lining which receives the image, is curved in man and in many higher animals but in the horse it is a flat and sloped surface, the slope being inclined backwards. This is the reason why the horse is unable to see objects above the level of his eyebrows.

This flat retina contributes to focussing difficulty. Another anatomical fact which contributes to this focussing difficulty is the poorly developed muscular body around the pupil. In many other higher forms of life this is quite strong and is able to pull or release the lens so that it may change in thickness and so act in focussing. By head movement, however, some of these focussing problems are overcome. If the horse raises his head he can see things at a distance much better, although his long range vision is relatively poor. If he lowers his head, close objects come into sharper view. By moving his head in a lateral direction, to one side or the other, the horse is capable of making positive use of the relatively flat retina for focussing.

The Highland pony, living naturally outdoors throughout the year on the rough terrain of the Highlands where good grazing would exist only in isolated portions of land, required to have an improved type of close and forward vision. Some breeds, notably the Arabian, have prominent eyes which allow them to have better visual scope than the majority of horses. But it appears that the Highland pony, unable to possess the luxury of very prominent eyes whilst feeding in deep heather or bracken, has improved its close forward vision in the course of his adaptation to his native land by changes taking place in the set of the eyes, the typical shape of his face and the set of his head. The Highland pony has his eyes set over a slightly narrow nose bridge allowing the eyes effective movement more further forward and closer than many other breeds. The slightly narrower nose between the eyes in this breed facilitates forward binocular vision better than in some others. Of course, this shape of the head, with its expansive forehead and set of the eye, has long been considered to be a sign of intelligence in horses generally.

Justification in this old belief of intelligence in horses with this type of eye-set certainly exists. Such horses have much better frontal vision than others. As a result of this they are able to focus better; they have improved images on their eyes; they react better within their environments; they can also perform certain tasks that require good eyesight, better than other horses might manage. Horses with an adequate degree of frontal vision, in fact, are able to utilise binocular vision, a valuable asset in adaptation. The Highland pony undoubtedly is capable of utilising binocular vision. This is the kind of visual property which gives the animal the advantageous ability to focus quickly on objects directly in front of him. The horse with this capability is obviously at an immediate advantage in dealing with mountainous complicated terrain. Presented with dispersal of good grazing in ground with numerous small obstacles such as rocks and marshes, binocular vision can permit highly efficient selection of nutritious herbage. The ability to cope with such characteristic efficiency in the face of these problems certainly contributes to the Highland pony's merited reputation of intelligence.

Best forward or frontal vision is obtained when the horse puts his face perpendicular to the ground, that is to say, when he flexes his head at the poll. Poll flexion is a common form of head carriage seen in the Highland pony, particularly when he is on the move and when he is under saddle. This gives him the best possible vision of the ground immediately before him. At his most usual rates of travel this is a principal need among all his sensory faculties. When the horse is required to see objects at a greater distance, he is obliged to raise his head. When he does raise his head for this purpose he loses both side and ground vision.

When the behaviour of Highland ponies at mountain grazing is studied at close quarters it is readily seen that they employ frontal vision to a substantial degree. This ability is also of value to those animals on smaller islands who have also acquired the habit of going onto foreshores in order to forage for scarce food among seaweeds as a form of foodstuff. The adaptation for forward vision has had its consequences. The principal one is the development of a hair screen over the eyes with the long forelock so typical of this pony. This is essential to deal with wind and driving rain, while continuing to graze, in horses that have their eyes directed slightly more forward than to the side. Another development which has gradually taken place in the Highland pony, to facilitate his ability to focus his vision at ground level while seeking out scarce food, has been a change in the location of the orbits in the head.

By examining the profile location of the eyes of typical Highland ponies, disregarding binocular frontal vision, it has been found that there is a fixed characteristic. A profile view of the pony's head allows a line to be drawn from the ear base to the nostril. The location of the slit of the eye on that line down the profile is found to be fairly constant in better specimens of the Highland pony. This location is approximately 40% of the way down this line. This gives a ratio of forehead to face in the order of 4:6. This means that the Highland pony's eyes are set further down his head, or further down towards his face, than in many other draught breeds, improving close-to-ground vision. Another consequence of this is that the pony's forehead, taken as the horizontal lines between the eyes and between the ear entrances, has become relatively larger than in some other breeds. A scientific examination of the distances between the lines between the ears, the lines between the eyes and the lines between the nostrils has been made using photographs of profiles.

Taking the relative proportions of these distances down the profile we find breeds of horses which have head characteristics, or head dimensions, essentially similar to the Highland pony. Such a study on 62 breeds of horses of the major types which can be found in the world, but particularly throughout Europe, showed that those horse breeds which were found to have head and face dimensions similar to the Highland pony included several of the other British breeds such as the Dartmoor, the Dale, the Fell, the Welsh cob, the Hackney and the Exmoor. Two Spanish breeds were also found to have the same forehead to face ratio. These are the Kladruber, and the Andalusian. Yet another breed, with the same forehead to face construction, is the Konik breed which is a native breed in Poland and Eastern Europe. The Norwegian and the Fjord horses of Norway also have essentially similar forehead and face dimensions. Again, the Finnish breed of horse has a head of the same category. Four Russian breeds have been found to be in the same class of head type also; these are the Orlov, the Lokai, the Karabair and the Don. In addition, the genetic links between the Highlander and Celtic, Roman and Nordic strains, together with certain Northern Russian strains, are evident. The genetic association with the two Spanish breeds is also very strongly indicated.

From this evidence alone there is justification for the historical origins of the Highland pony which have already been suggested. These head characteristics show that the Highland pony, together with certain other British pony breeds, has acquired facial characteristics from the heredity of similar environmental challenges in complex and rough terrain.

When the Highland pony is allowed to select its own head position he knows instinctively how to move his head and neck in cantilever fashion to ensure that he will not stumble over obstacles at ground level. The head characteristics, including the set of the eye and the carriage of the head with flexion at the poll, which are a feature of this breed, make it unnecessary for the animal's head to be held down with martingale for the purposes of riding. This gives the animal freer head movement. All of this contributes to its functional characteristic of surefootedness. When being used over rough ground for riding this asset makes the garron a breed of choice.

Another environmental adaptation is the length of mane and the length and density of the hair of the tail. In severe weather when a high wind is blowing the Highland pony, when he ceases to graze, will normally direct the hindquarters into the wind. This allows his tail to be blown between his hind legs while he holds his tail in close to his dock. The tail hair then shields all of that hairless area of his perineum, his inguinal region and the inner thighs. By this means he spares himself the loss of heat and energy reserves, essential to take him not only through the winter but also through a spring that can be long and harsh in some years. The combination of cold and wind, especially in wet weather, can inflict great stress upon a horse. Nothing so chills a horse as keen winds with rain or snow. The quantity and quality of hair over the head can both waterproof and watershed, to ensure that ears and eyes are not subject to weathering. The effects of frost in very cold weather can be worst in a saturated horse. This garron mane, with its unique length and density, can act as a waterproof screen for the head, jowls, throat and neck. When left uncombed numerous long, ropey segments become formed throughout the mane of Highland ponies. Because of the inherent wavy nature of this hair it has the propensity to intertwine and to "plait" naturally.

Apart from the eyes and ears, these anatomical areas of the head and neck contain vital superficial structures, which require heat conservation in winter. The large, long and superficial jugular veins, and their upper tributaries in the throat and about the head, can be readily cooled and so reduce the temperature of the blood as it passes through them. (This fact is used to advantage in treating heat stress in the horse by ice-packing or water-hosing the jugular grooves). The sensitive tissues and glands in the vicinity of the throat also require thermal protection.

The density of winter coat and the direction of hair lie, especially over the hindquarters and back, provide an outer periphery to the body which acts as a weather shield, so efficient that ice may form on top of it without transmitting chill to the skin.

The feather, on the fetlocks of the Highlander, is just sufficiently developed to divert and take the run-off of moisture being channelled down the legs to the fetlock points. This saves the hollows and bulbs of the heels—and to some extent the frogs of the hooves—from excessive wetting and the undesirable consequences of this such as a predisposition to thrush locally and to chilling generally.

The hirsute nature of forelock, mane and tail in this horse gives it a considerable amount of protection in the summer when the

Highlands and Islands are plagued by midges and biting horse-flies or clegs. These insects collectively represent an environmental feature so adverse as to reduce habitability in this region when climatic features are otherwise at their most benign. Other forms of livestock and man can sometimes become frenzied by this irritant fly-life but the pony has more tolerance of it. This is due largely to an ability to maintain a defensive fly-screen by continued gentle agitation of its long hair, fore and aft.

Variability of size in this breed has been a constant source of discord among breeders for many years. In the way of concerned mankind, they have tried to arrange order out of apparent disorder and have arbitrarily subdivided the breed by size into a larger "Mainland" and a smaller "Western Isles" type. These types in turn are considered to represent the working or draught type and the riding type respectively. This conception is on one hand not unreasonable, but on the other hand naive, if left at that. The insular families of pony, particularly on the smaller islands, are more subjected to onshore wind, whilst at the same time deprived of shelter in a tree-less terrain. Their thermo-regulatory adaptation to this is a reduction in physical size, including limb length. This is a proficient means of body-heat conservation. The larger islands offer more potential shelter against driving wind and such adaptation is not a high priority in pony life there. The Isle of Skye ponies, for example, partly by selected breeding and partly by tolerant conditions, are often among the larger specimens, over 14 hands. On the mainland, some Clydesdale influence may have contributed to size in some individuals of half a century and more ago, but the myth of mainland type as being more suitable for draught work than riding is difficult to prove. Versatility could be the Highland pony's middle name, for it can perform between shafts and under saddle equally well, irrespective of height and coastal classification.

That some insular adaptations have occurred is without question. For example, the Eriskay pony, whilst keeping strength, is of recognisable fine build. A society now exists to work for the preservation of its type in pure-bred form. The ponies of the Isle of Rhum have been notably insular in their breeding. They also have very recognisable characteristics, the most obvious being growth of hair. Inter-island traffic of horses has long been adequate to create a well-mixed population, one which overlapped with the mainland population. This trafficking has dispersed local adaptations, contributing most of these to the genetic pool of the breed as a whole. The dictates of controlled breeding, whether based on foresight or fashion, wisdom or whim, can still operate, however, to draw selectively from this pool and modify representatives of the breed. Such is the effect of domestication.

The main features of Highland pony conformation are good development of fore quarters, hind quarters and neck and a compact general appearance. The strength of the fore quarters has been particularly emphasised. Some outstanding specific features of the conformation of the Highland pony include a well-set neck, one which is particularly well developed at the base. The breast also has been described as being typically well developed and wide. The shoulders are described as being well muscled and steep, or oblique, in direction. The hind quarters are visibly muscular and the conformation of the loins is compact. Shortness of cannon bone is a preferred feature.

All of these features are developments which have taken place in the breed as a result of its adaptation to its work requirement. Doubtless this adaptation has been guided by selection through breeding. The total result of all of these physical features is strength and power, making the animal particularly suitable for endurance riding and draught work. The draught work called for is in various forms. A conventional form is traction of carts, ploughs or other agricultural equipment. Often this has to be done without the assistance of a second horse. Another form of draught which this animal is called upon to provide is porterage. Highlanders originally used the native pony to assist them to carry peat in panniers and other items of domestic or personal requirement slung across the back. Its use in deer stalking represents very exacting draught work. Much of its typical work is over ground where there are no roads or tracks, and ground which is beset with obstacles such as rivers, bogs, marsh and rock, and thick heather.

As a consequence of all these requirements, the breed has become physically suited for miscellaneous draught work. The animal has also acquired the physical capacity to utilise leverage, particularly on the fore quarters. Such leverage capacity is essential in assisting the animal to alter its centre of gravity in relation to the dead load whilst it is negotiating difficult terrain. The leverage mechanisms involved are interesting. The principal leverage mechanism is a very well-developed neck with a strong shoulder joint. In any horse the neck is the instrument of leverage. Apart from the development of the neck—and the particular development of the neck at its base where it blends in strongly with the shoulder—there is also the development of another physical mechanism to assist leverage. This secondary mechanism is the ratio of forearm to cannonbone, where the forearm is comparatively long and the cannonbone is comparatively short. This gives a high ratio of forearm to cannon. Such a high ratio of forearm, which is muscular, to cannon, which contains only tendons (extensions of muscles), gives an efficient pulley system and allows greater leverage of the limb.

As a consequence of these special characteristics certain groups of muscles in this breed have become particularly well developed. These are the muscles of the neck, the shoulder area and the forearm. These muscles are particularly involved with the sling arrangement by which the forelimb functions against the body—the forelimb, of course, has no bony or joint attachment with the fore quarters. Various muscles have specific jobs but in general they serve the purpose of drawing the broad flat shoulder blade forward and backward whilst the limb advances, takes weight, then pushes forward. In the course of this type of forced movement of the forelimb, the angles of the shoulder and elbow joints alter very little. During the course of progression of the forelimb, the angle of the shoulder in particular is held fairly constant. The greatest amount of movement that takes place in the bones of the forelimb occurs at the height of the scapula—in the region of the withers—and below the carpus or knee. Since these are the two parts of the forelimb required to alter position most in effecting leverage we can understand why the musculature relating to these parts has become particularly well developed in this breed.

The musculature involved in flexing the knee and in causing the cannon bone to be used as a lever are principally the extensor and flexor. The extensor muscle is the large well developed muscle which runs down the front of the forearm to the knee. The flexor muscle is the strong one which runs down the back of the forearm from beneath the elbow towards the knee joint. While these are the major muscles involved in this type of action there are supportive muscles which do not require special identification in appreciating the whole principle involved in the work of forelimb leverage in draughtwork.

Another feature of the action of the body in draughtwork in both of its forms of traction and deadweight carriage requires strong muscular exertion in the region of the loins together with the obvious need to have strong musculature in the hind quarters. It should be borne in mind that when being extended to drive the animal forward, even against considerable weight in draughtwork, the thighbone or femur never extends beyond a vertical line from the hip joint to the stifle joint. As with the forelimb, a ratio of gaskin (or second thigh) to hind cannon, in which the latter is of minor length, gives the best coupling arrangement to effect hind limb leverage. For this reason the description of the ideal conformation of the hock of the Highland pony as being "well set down," has good functional justification. A well set down hock with adequate musculature in the region of the second thigh gives maximum potential to the limb to extend the cannon segment of the leg and improve not only strength of support but leverage.

In order to maintain dead weight over the back, the back itself requires to be comparatively short. For this reason the best conformation for this breed has been described as a back that is medium in length. Additional length in the back is likely to contribute to muscular stress in supporting weight. Probably the most critical part of the horse's anatomy in maintaining substantial amounts of weight on its back are the loins. It is essential in this animal, and a fairly constant feature also, that the loins be strong and well coupled.

The slope from the croup to the tail head in this animal is sometimes seen to be comparatively long and steeper than in some other breeds. This is not thought to be an attractive physical feature and for this reason there is sometimes discrimination against it in the show ring. Nevertheless, this type of physical shape of the hind quarters is often found in individual animals which have a particular strength in the hind quarters. It may be of passing interest to note that, although it was no draught animal, the thoroughbred Secretariat was almost goose-rumped in appearance because of this type of conformation. Secretariat is considered by many to have been the most powerful running thoroughbred in history. Evidence like this cannot be ignored.

Whilst levering the fore quarters in draughtwork the animal requires to follow-through with the forefoot and this requires a considerable amount of flexion of the fetlock joint. For this reason one looks for a good angle between the cannonbone and the slope from the fetlock to the toe. This angle has been described as one of approximately 45°. Such angulation would allow the forelimb to be carried through in vigorous action and also allow it to be well flexed for its return to a forward position. In the hind limb there is a similar follow-through property in action and for this reason the angle between the hind cannonbone and the slope of the pastern to the toe is also well developed. It is normally found in satisfactory conformation to be approximately 55° in angulation.

Yet another feature of conformation which requires to be well developed for traction work is breadth between the shoulder joints. Narrow-chested animals do not have the same traction power as those which are broad in the breast. Furthermore, the breast muscles themselves are much better developed and much more prominent in the stronger animal which is capable of impressive traction work.

In the main, these features which have been analysed above are those which are characteristically found in the so-called mainland type of Highland pony. They are less obvious in the so-called Western Isles type of pony. The latter is commonly referred to as a type more suited for riding. Certainly the conformation requirements for riding are somewhat different. There is still the need for the strength in porterage of course to be retained in the conformation of this breed, and for this reason one still looks for strength of loins, strength of hind quarters, an appropriate length of back and again development in the region of the shoulders and neck to permit leverage. It has already been emphasised that leverage is essential in altering the location of the load on the animal's back in relation to its centre of gravity, which passes through the body of the animal in line with the girth. For this reason approximately 60% of the animal's total weight is carried on its forefeet with the remaining 40% carried on the hind feet when the animal is not being ridden and is standing in a normal position at rest. This also is the reason why, in the ideal conformation, the hooves of the forefeet should be more expansive than those in the hind feet. In order to ensure that this amount of weight is

Alert frontal vision through a long
screen of forelock

An old Highland pony with
well-wrapped jowls

Muscular system
showing the triceps
(after Stubbs)

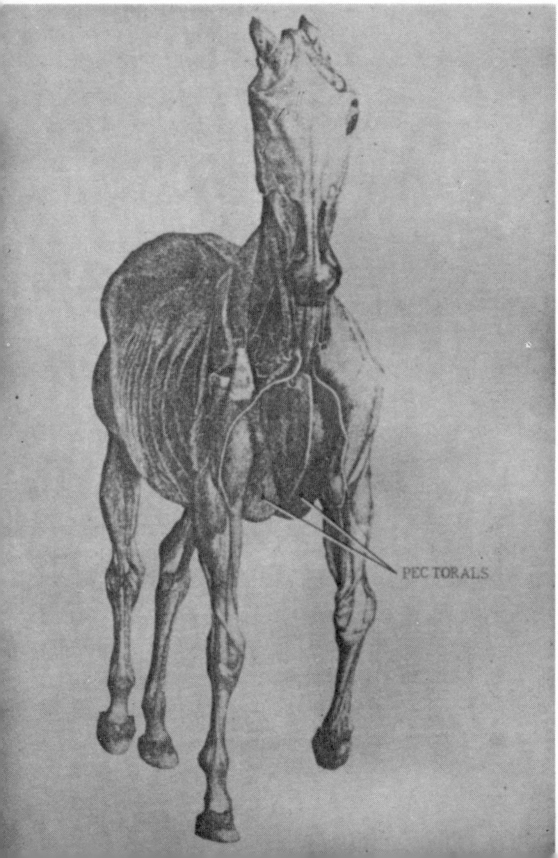

Muscular system
showing the pectorals
(after Stubbs)

Diagram showing the manner of weight bearing down on the chest and through the front leg, indicating the need for a bulging triceps muscle acting as a shock absorber

Ponies at work in a hayfield on a Highland croft *(Landseer)*

Ponies carting in tandem

Gathering peat in a slide cart

suitably distributed over this comparatively small surface, it is best that the hoof should be "open and rounded" in appearance. It follows from this that it is a fault for the toe of the hoof to be deviated from midline. An animal which is pigeon-toed or hen-toed cannot exercise a good follow-through action with its foot whilst driving against a load, for this undermines the entire function of leverage. It is obvious, therefore, that defects in the toe can represent not simply minor conformational faults but major functional defects.

The muscular activity taking place in the fore quarters and limb of the horse during its forward movement is a beautiful piece of mechanism. While the forelimb is moving forward and back the cervical and the thoracic muscles are alternatively contracting and relaxing. These are the muscles which form a sling in which the chest hangs between the two forelegs. When the leg is moving forward the very long muscle which extends down the side of the neck to the arm contracts and relaxes also. When it contracts it pulls the forelimb forward. The muscles of the breast also pass from the chest to the arm to aid in its leverage. The alternating strong contraction and relaxation of these muscles in pairs is known as reciprocal muscle action. It is this important function of muscles, anywhere in the major muscular parts of the body, which is essential in maximum muscular proficiency. Needless to say it is this same reciprocal muscle action which, carried out most proficiently, gives the animal not only strength, but an even and flowing gait.

The large, nicely rounded and prominent muscle mass about the elbow joint, which is a prominent feature of Highland conformation, represents the triceps muscle. It would appear that one important function of this muscle, which has not always been emphasised, is that it acts as a shock absorber mechanism for impact of the limb upon the ground, when considerable weight is put on the forelimb. Most of the body weight of the animal, together with its load, is applied to the foreleg. The tendency of such considerable force will be to force the elbow joint into flexion and this is resisted by the powerful mass of this triceps muscle.

There are other forms of special apparatus to take up impact, one being the suspensory ligament, the other a check ligament, both of which pass down the leg together with the flexor tendon, which also serves to reinforce the limb when the foot is pressed on the ground.

Behavioural evolution took place parallel to the physical changes. The end points of behaviour evolution are there to be recognised in peculiarities of vision, reflex actions, sensitivity to ground vibrations, flight, retentive memory, gregarious habits, the senses of hearing and smell. Adjustments to vision need changes of head position. As a result, horses have poor vision, most especially in the early days of life. These visual features are not handicaps in their own special habitats but they are serious defects in various domesticated situations. Young horses run into structural obstacles around farm premises; any horse may "spook" at minor innocuous factors suddenly coming into their vision; horses are often reluctant to load into trailers which have "unseen" interiors.

Good prompt reflexes to potential danger allow horses to be reactive to novel stimuli. This may take the form of flight, such as bolting.

The horse, as it matures, acquires a very good memory. This memory seems to be selective for unpleasant events but a well-trained horse remembers its training all of its life. This good memory tends to fix habits, both good and bad, and some observers emphasize the latter.

The herding instinct of the horse gives it a range of gregarious forms of behaviour. Gregariousness facilitates shared learning between and among horses and also is the basis of obedience to the herd. Close herding is noticeable in Highland ponies and they are an excellent choice for group riding. Foot sensitivity in horses makes many of them ultra cautious about footing. Horses tend to avoid rushing into water since underwater obstacles such as stones are unseen. A complaint against some Highland ponies' reluctance to go into water or stand in moisture is only partially justified as a criticism. It is behavioural wisdom and, when it occurs, it is also evidence of antiquity. Skin sensitivity is keenest in the region of the garron's lower flank, making it sensitive to the crop or making it appear ticklish there. The sense of hearing in the horse is often good. This is essential for appropriate behaviour in bond formation to be encouraged by vocalisations between breeding mares and stallions, between brood mares and foals. The range of equine hearing exceeds the human range and the ear mobility of the horse, being so great, enhances this sense. Harsh noise can be very inducive to anxious behaviour in the horse. The lack of harsh noise in the Highland environment must surely reduce anxious tendencies.

Horses usually have a very good sense of smell. They may respond adversely to strange unnatural odours. The sense of smell helps in the search for scarce food. No urban alien odours normally concern the Highland pony in its formative years and anxious elements are not introduced to its behaviour by this type of experience.

The many claims of the excellent temperament of the Highland are evidently justifiable but these are not totally heritable, and environmental influences in the most formative age up to two to three years are important. For this reason the Highlands and Islands will continue to be the best nursery for the breed which can be recognised as Caledonia's own.

Chapter V

THE JACOBITE TIME

Robert the Bruce held the Scottish throne capably, but following his reign the crown was in frequent jeopardy from covetous noblemen. In 1390 Robert II died, and was succeeded by his son Robert III who was in poor health and unsuited to the throne. The heir was then his son James, and plots on the life of James were rumoured. He was sent to France for safety. In the voyage he was captured, taken to London and held a hostage by King Henry IV. The trauma of this killed his father and James was now proclaimed King James I of Scotland. He was destined to spend the next eighteen years of his reign an English captive, and his throne consequently lost most of its authority.

Major families throughout Scotland began to assume responsibility for the administration of laws within their own areas. As these families acquired land which they were able to assign to others for farming, they also acquired power to raise forces within their territories. This parochial system of relative independence became the pattern throughout Scotland. In the Northwest, the Chiefs of families known as clans acquired similar power. They became styled as Lords and enjoyed the status of autonomous monarchs, aggressively disregarding the existence of a central government in Scotland.

Throughout the Highlands the clan system became the fabric and design of Gaeldom. Life had been much like this for the preceding century, but the loss of power in the capital city of Edinburgh reinforced the system in the North of Scotland. In the Highlands the influence of both the Church and the State was minimal. In this system very different loyalties prevailed and varying standards were observed. The Clan Chief was a paternal figure who virtually had power of life and death over the clan members whom he commanded by obligation to him for land with a style of absolute authority. The loyalty of the clansmen to the Chief was absolute. The clan system was not feudal. It was a patriarchal social system which had a Celtic concept.

For members of the clan the important facts of life related to the protection of home, the cultivation of marginal land, and to the conservation of livestock. Protection of their animals was a matter of great concern. The kinship of the clan afforded protection of this kind to ensure that the raiding of animals by opportunists would not be profitable. This mode of self-defence was also the means whereby clan chiefs could, on very short notice, raise a respectable size of army from the ranks of his own clansmen. For the most part the clans held each other in healthy respect. Boundary feuding existed, but it was not excessive. Thus the Highlands of Scotland with the clan system had forces of reserve militia awaiting any event, for organisation and activation.

As the Highland pony entered the 18th century, which was to become a very important time in its history, it was genetically well equipped. This was a testing time, the century of its consolidation. It was uniform in type now and featured very many desirable characteristics, both physical and in temperament. Prominent features included the immediate needs such as hardiness to withstand the rigorous climate of the north and west, strength for draughtwork, stamina for long spells of toil, docility and a compliant temperament.

By now, life in the Highlands, whilst not wealthy in material terms, was nevertheless rich in its social pattern and in self-sufficient customs. The form of micro-farming, which we refer to as crofting, required horse power of a very appropriate type. It was no mere coincidence that the appropriate type was to hand. The work on these small portions of arable land very often required that a single horse be used. The need was for a horse of medium build but of great strength to cope with the rocky subsoil; a horse so highly domesticated that it could be used by man, woman or child, with complete safety and proficiency; a horse so versatile in its uses that there was no task for which it was not capable or of enormous value.

The chores of the Highland pony within this culture were extremely varied. Peat was the form of fuel used in cooking and in heating the rough habitations. This peat required to be dug in bogs quite often far removed from habitation. Pony power was required to transport these peats from these bogs, across stony lands to the homesteads in order to supply domestic fuel the year round.

In the springtime the animals would be required for ploughing. Arable land was often in confined areas with intricate contours—a thin subsoil over rock. Seaweed, as fertiliser, required to be carted from the shore at low tide. Subsequent harrowing of ploughed land again would require the peculiar skills of the Highland pony—stamina, courage, power, manoeuvrability and compliance. At harvest the cutting of hay and of corn did not require horse-drawn machinery, as a rule, since the cutting was done by hand. Nevertheless, horses were used to draw carts on which hay and corn would be carried to the stackyard close by the house for convenience during the winter for feeding livestock. Some of the draughtwork of these animals was done using panniers of wickerwork, draped one on either side of the animal. Sometimes the vehicles being drawn by the animals were wheeled carts, sometimes they would be sledges. On other occasions they would be a very peculiar type of wheel-less cart or slide car consisting of two long poles which were dragged behind the horse along the ground; across these poles would be built a hod with a floor and a back of wood sufficient to carry supplies of log and peat. Occasionally the horses would be used to draw carriages of families who were visiting far from home, families who were attending some special social event. They provided regular transportation for the more mobile people in the community—physicians, traders, landowners. The horses were also used for riding, but by and large they were not used for this purpose excessively. In the main the crofters conserved the horse and saved its legs as much as possible. They chose to walk with their animals, rather than ride them over rough ground. When they were ridden they were generally taken at a slow pace. Again this was an effort to extend the life of the animal as a member of the work force. By no means was this any indication that the animal did not make a good mount or have the ability to move with a free style. In fact, this animal has long had very distinguished gaits.

Although stallions were certainly docile enough throughout most of the year to be hand-led by almost anyone, their breeding value was recognised to be of critical importance and their work was usually limited to this. Mares were used for breeding for much of the time and would be occasionally used for riding. They would certainly be used in harvest work.

The ponies which did most of the more strenuous work were

the geldings. The Gaelic word to describe a castrated horse is garron. For this reason, the garron was the most common form of word for the Highland pony. The name in due course became applied to the breed in general, particularly by non-Gaelic speaking people who came to recognise the breed and came to appreciate it. No other breed of horse draws its common name from its emasculated males—surely a unique compliment to this breed, that even its castrates deserved and won such wide recognition and such a seal of approval. The emasculation, or cutting, of the males was particularly necessary for the demands of spring work. Using the light of day characteristics for its biological calendar, the Highland pony has probably a shorter breeding season than ponies indigenous to lighter, more southern latitudes in Britain and elsewhere. For this reason there is a more defined breeding season, in the late spring of the year. While this breeding activity is concentrated, entire ponies are extremely active and alerted for breeding. With a little stretch of the imagination, their behaviour at this time can be seen to resemble the rut which accompanies male breeding in certain other animals, such as deer, in the same latitudes. Throughout the remainder of the year, the sex drive of the entire Highland pony presents no real trouble. If they retain one male characteristic throughout this time it is that they have less patience and tolerance of chronic tedium than their castrated brothers, the true garrons.

The work of the garron did not end with crofting. He was an animal frequently used by cattle drovers. The drovers were a very distinguished group of cattlemen in the Highlands. They were given special privileges in law dating from 1725, relating to their right to carry weapons and to wear tartan, the national dress which was banned for decades in Scotland by the British Government following Culloden in 1746. Young men who had been allowed into the fraternity of drovers were considered to have risen in society. Native Gaelic speakers, they were of course required to speak in English since their work mainly concerned the transportation of large numbers of cattle on the hoof from the north of Scotland down into England. This journeying also brought them into contact with a wider spectrum of society in the course of which they acquired some of the more refined manners and habits of the gentry in cities and townships to the south. Mounted drovers wore trews. These long, tartan, close-fitting trousers of proper Highland riding habit became popularised in southern Scotland. There they have been worn as uniform by Scottish regiments since. On their return from season-long drives of cattle into England, drovers would be welcomed in their home townships where they would entertain their relations and neighbours with tales of their travels and news from the south. Their influence on society was substantial. Garrons undoubtedly helped their image.

While Scotland's social scene seemed secure, and trading relations with England were amicable enough, disastrous events were brewing in politics and at court. In Edinburgh the Scottish throne had passed down in due course to Mary Queen of Scots, whose tragic life is well known. Her son was James VI, and, on the death of Queen Elizabeth in England, he became James I of Great Britain. The Union of the Crowns was thus effected in the year 1603. Following some political confusion and disagreement, the Union of Parliaments between Scotland and England followed in the year 1707,

The century between the two events was a most troubled one for the people of Scotland. The Covenanters, who strove and battled for the privilege of pursuing their own religion, were involved in many struggles and much bloodshed. Oliver Cromwell, who had overthrown the English monarchy for a period, arrived in Scotland with an army which destroyed many Scottish fortresses in the South. At this time he was established as Lord Protector of England and of Scotland and even the Highlands were temporarily subdued under his influence. In 1658 Cromwell died and his Protectorate soon collapsed. The former King Charles, who engaged in numerous adventures and had managed to evade Cromwell's net in making his escape to the continent, was restored to the throne. In 1685 he was succeeded by his brother James VII who in due course fled to France before the forces supporting his Dutch son-in-law William of Orange. William and his wife Mary were proclaimed King and Queen of England, Ireland and Scotland in 1689. In the North of Scotland, however, many of the people remained true in their allegiance to King James in exile. They became known as Jacobites. Immediately a confrontation took place between them and William. A body of troops was sent to put down the Highland Jacobites. As these troops passed through the narrow pass of Killiecrankie in Perthshire, the Highlanders who had mustered to meet this force attacked it in the gorge and defeated it entirely. The Jacobites, however, lost their leader in the battle. The Highlanders returned to their hills and glens.

This disturbance in 1689 was an indication of the shape of things to come. William's reign saw the massacre of Glencoe. The Government had issued a proclamation ordering the clan chiefs to take their oath of allegiance to King William no later than January 1, 1692. If they failed in this they were to be put to the sword. At this time of the year it was, of course, the depth of winter in the Highlands. The branch of the clan Macdonald which occupied Glencoe was led by an elderly chief who due to age, oversight and the severity of weather arrived three days late at the office of the Sheriff of Inverary to take the oath. For his failure to meet the deadline, the government decided to take punitive action against his clan. The massacre in the pass of Glencoe resulted.

Glencoe did nothing to improve William's acceptance in the Highlands and disillusionment with the Dutch monarch increased. At this time Scotland was experiencing a long period of severe economic depression at the end of which Scotland as a nation was virtually bankrupt.

Meanwhile in exile James VII had a son called James Edward who in the view of many was the legal heir to the throne, but in 1701 an Act was passed through the English Parliament declaring that the English throne in due course should not pass to him. William of Orange died in 1702 and was succeeded by his sister-in-law Anne. The Act specified that on Anne's death the English crown should not go to James Edward Stewart but to a German princess, the Electress Sophia of Hanover. These events continued to generate Jacobite feeling, particularly in the North of Scotland. In an effort to recognise this, the Scottish Parliament passed its own Act which specified that the successor to the throne should be a descendant of the House of Stuart and not Hanover. The Jacobite cause was not only acquiring sympathy, but its justification was now being acknowledged by Act of Parliament in Edinburgh. This Act was superseded, however, by the Treaty of Union between the two parliaments. The Jacobites, as might have been expected, totally rejected the Treaty. They spoke out against it and in favour of their "Ancient Mother Caledonia." Nevertheless, the Scottish Parliament was dissolved and consequently Scotland lost its independence, not on any battlefield, but as an anticlimax in its own debating chamber in Edinburgh. When James Edward Stuart's father died in 1701 he was promised help by the French to re-establish his right to the British throne, and in due course this was attempted. This first Jacobite rising started with considerable chance of success to the great concern of King George, now monarch in England.

The Jacobite Rising of 1715 was planned to take place from three points within Britain. One was to be in the Bristol area in the Southwest of England, another was to be located in the North of England, while the third was to originate in the Highlands. The message that went out to the Jacobites in the North of Scotland to announce the plans for the uprising was quite forthright. Among the instructions which were issued to landowners and tenants was the following statement: "You are to tell the gentlemen that I'll expect them in their best accoutrements, on horseback and no excuse to be accepted of."

The uprising in the Southwest and in the North of England lacked tenacity of purpose and did not really succeed in becoming properly established. The uprising in Scotland, however, was real. Leader of the Scottish uprising was the Earl of Mar. The clan system allowed Mar to raise his forces quickly, and a substantial army was soon mustered. Among that army was a force of mounted men. They could not be called trained cavalry in the customary military sense but they were a force to be reckoned with and served as a valuable defence for the main force of foot soldiers, a force of some 5000.

At first this Highland force outnumbered the Government troops but the latter were substantially reinforced very quickly and in the subsequent encounters it was found that the Government dragoons mounted on cavalry chargers were an uneven match for the gentry, as they were called, riding on Highland ponies. Nevertheless the Jacobite forces created very substantial opposition to the Government troops, winning numerous engagements. Many of these took place in the vicinity of Edinburgh, Stirling and Perth. In the encounters, historical descriptions of Mar's mounted forces report that though they were brave, enthusiastic and mobile, they lacked military training. This was to be a serious defect in their battle schemes in general.

Mar's cavalry, as it could be called, initially numbered about 600.

As the uprising progressed, the force came to number a thousand mounted Jacobites. Horses were almost entirely drawn from Highland sources including the Hebrides where Spanish horses had recently been introduced by Clanranald, chiefly in South Uist, but in adjacent islands also. For a period the infantry and cavalry together were a match for any Government army which could be sent against them. Indeed the Government was never able to muster a force of cavalry which exceeded in number that of the Jacobites. At the battle of Sheriffmuir, the Chief of Clanranald was killed. He was a great loss, having been a colonel in cavalry service in Spain and experienced in battle. The uprising progressively lost its momentum, although for a while it had the support of King James himself. He had eventually arrived from Europe to take his place with the Jacobite army. James apparently had no taste for battle and remained only a short time in Scotland before he slipped away, very unexpectedly, in a ship bound for the continent, from which he never returned.

The continuous pressure of the Government forces caused the Jacobite organisation to become disunited. The enthusiasm had gone out of the cause and zeal for the uprising was spent after the King and Mar had fled the country. Thus in due course the uprising terminated and by May 1716 all was quiet. The disenchanted clans had laid down their arms and their leaders were either in hiding or were obliged to flee overseas. Many Jacobite prisoners were taken and dealt with harshly, some executed, others dealt with more leniently and allowed to settle abroad or to return to their homeland. There are many "ifs" in the history of this early Jacobite rising. It might have succeeded had certain events been different. These "ifs" allowed the fabric to remain in Highland society of a Jacobite ideal which was soon to grow again in another rebellion.

An attempt was made to disarm the clans totally, but this was not successful and their weapons were concealed to fight another day. The Union of Parliaments was now more unpopular than ever and even those Scots who had been till then impartial began to resent the Government's attitude to their countrymen. King George in London directed that the policing of the Highlands by his militia be improved. For this purpose he appointed General Wade as the English Commander-in-Chief in Scotland. Wade embarked on a programme of military road-making to create a network of communications through the otherwise impenetrable regions of the Highlands, linking and strengthening the strategic fortresses at Fort William, Fort Augustus and Fort George. In this way the Highlands were opened up in a way that they had not previously experienced, and to some extent this eroded the nature of the indigenous social system. In 1727 George died, to be succeeded by his son George II, but the attitude between the monarch in London and the people in Scotland did not change. Scottish resentment of taxation on various imported goods was strong and widespread, and the smuggling of landed goods inland by pony came to be recognised as a legitimate exercise for many. Smugglers became local heroes, but were subject to execution. Meanwhile, King James in exile in Europe now had a son, Charles Edward Stuart. In him the Jacobites saw fresh hope for their cause which had endured so many decades of frustration and disappointment.

The early descriptions of the young Prince Charles reaching Scotland were that he was a young man of great courage, energy and considerable personal magnetism. This was a figure the warlike Jacobite people could readily accept in a way they found difficult with his less attractive father.

When young Prince Charles left his father in Rome and set out for France in 1744 he was, of course, fully informed of the limited character of the support that existed for his cause in Britain generally. For reliable support he knew he would have to look to Scotland, and to the Highlands in particular. The filtered news which reached him encouraged him to take positive action, and he decided to sail to the Hebrides to seek support for his claim to the British throne.

In August 1745 Prince Charles, with seven supporters, landed on Eriskay, one of the smaller islands by South Uist in the Outer Hebrides, where he quickly acquired intelligence reports about the potential forces that he could raise. These were not encouraging. He wasted no time there and set sail again for the mainland coast of Scotland and two days later anchored near Moidart. He then established himself at a base near Borrodale in Arisaig. From there he sent out appeals to clan chieftains and in particular to Cameron of Lochiel, the head of the Cameron clan. Lochiel and other leaders were not convinced that the time was right for a national uprising on behalf of Charles. Prince Charlie's personal charm became evident. Although he was only 24 years of age, he swiftly imposed his will upon Lochiel and Clanranald and not only won them over, but won an oath that each would muster every man able to carry arms within his domain. Once he had this assurance, Charles went on to Glenfinnan where his standard was raised. This was to be the point where the first rallying clans would come to join forces with him. He himself had virtually no equipment and had landed with but a small quantity of military stores—just over 1000 swords and very little else. But the clansmen rallied to his banner in a day, and by evening he had raised 1200 men dedicated to his cause. Chieftains and other leading figures among the clans were mounted and Charles became equipped himself with a well-chosen Highland steed, doubtless with Spanish blood in it, from the South Uist importation by Clanranald 32 years before. This was the most favoured Highland horse of the time.

Charlie's force quickly marched on Edinburgh and camped outside it. A tapestry made at that time depicts the central part of this force kneeling in prayer before they embarked on their first engagement at Prestonpans. The tapestry includes Prince Charles mounted on a splendid light-coloured Highland stallion.* He wore long boots over his riding trews. Among the hundred Jacobites in this tapestry, five are horsemen. Cameron of Lochiel is mounted and two other Jacobite leaders are seen mounted and wearing tartan trews. Lord Nairn is seen on a dun coloured horse wearing a kilt over his riding trews. Some wore long boots over their trews. Charles sits on his mount as a trained equestrian readily identifiable as "The Young Chevalier," a popular name that he was to acquire at the same time in Edinburgh. The populace there, although initially cool to his cause, warmed to the man and to his charisma, and soon he won their complete affection. That he was an out-outstanding, confident, personable leader of men was in no doubt. The thousands of clansmen who were dedicating their lives to him would make no such gesture to anyone they deemed possessed of lesser qualities.

Charles mustered his infantry and horse forces in Edinburgh. The Jacobite force then marched eastward to meet General Cope with the Government army which had meanwhile landed at Dunbar.

* It is interesting to note that the steed on which the Prince is mounted in the tapestry is of a type now increasingly being sought after and being brought back to the leading show-rings in this country. The type is eminently suited for trekking and endurance riding whilst it still is an animal of considerable style and presence.

The Prince and his forces rested in a field close to the Government forces overnight. At 3 a.m. the following morning on the 21st of September they moved forward in the thick Scottish mist close to General Cope's redcoats. Cope had trained militia and 500 cavalry and he contemptuously dismissed the opposition as "a parcel of rabble, a small number of Highlanders, a parcel of brutes." As light rose on the day, the Highland force advanced three deep, broken into their small clan units, led by their chiefs. Their arms consisted of a targe or small arm shield on their left arm, a dirk in their left fist and a claymore in their right hand. They rushed forward, plunging into the ranks of the enemy. In eight minutes they had shattered the entire Government army. General Cope and his cavalry were the first to flee and galloped south for England. They were not pursued. Charlie's forces showed leniency in victory, giving mercy to the injured and captured. After this victory, the Highland army returned to the capital piping the tune "The King shall come into his own again." They were hailcd as heroes when they returned to parade down the High Street. Citizens of the capital—and the ladies not least—took to the Prince without reservation, and the romantic aura surrounding the Jacobite leader and his cause grew apace. This was to be its main recruiting force, its main source of inspiration to the bitter end of the fourteen month campaign. It also was to be the legend which survives it still.

Further Highland support was filtering down to him, and his mounted strength increased to 300 horse. Charlie then felt that he had sufficient forces to justify a march on England and set off south in the direction of the first English city, Carlisle. As the forces neared Carlisle the citizens of that town did not know what manner of army to expect. They went to the walls of the city to defend it, firing off muskets at sheep and other animals outside the city, such was the alarm and confusion which tales of barbarous Highlandmen had created in them. Carlisle was taken in majestic fashion: Charles, mounted on his grey Caledonian steed and supported with a hundred pipers, advanced upon the city. It immediately surrendered to him; the remainder of his forces then quickly occupied it and the second great victory had gone to the Jacobite cause. On this occasion barely an effective shot was fired. The count of the strength of the Highland forces was given at this time at 5000 foot and 600 Highland cavalry. It was no complete army but it was a substantial force to be reckoned with. It had high morale, being mustered in less than two months by the charm of the newest Jacobite pretender to the British throne. The force left Carlisle and marched south into England. Charles marched with his men, mingling with them easily. He had been an athletic young man in Europe and had tasted battle there. He was also a linguist and was quickly learning Gaelic, so was able to communicate with the Highlanders. He was one of them now.

Although the Government army could have been diverted to block his path on his march south this was not done, and the Jacobite forces marched down throughout England without real opposition. They finally reached the town of Derby on the 4th of December. At this point they were only 130 miles from London. Charles was then in command of a confident and unbeaten army of Highlanders devoted to him and his cause. Complete victory seemed at hand, with London apparently at his mercy. In the city there was civil panic. Shops closed, there was a run on the banks, King George himself was getting ready to flee the country and return to Germany, while members of his government were planning a hasty conversion to the Jacobite cause. At this, the most crucial moment in the campaign, his principal adviser, Lord George Murray, to the astonishment of the Prince, strongly advised the return of the army to Scotland. A bitter council of war took place and Murray seemed to be able to convey the logic of his planning to other clan chieftains, leaving Prince Charles isolated in his conviction that a judicious withdrawal was unnecessary at this stage in the campaign. Outvoted in the council of war, he grudgingly accepted its decision to return to Scotland. On the march back much of the excitement and the enthusiasm had gone out of the adventure, but the Government army continued to avoid the rebel army. In due course they returned to Scottish soil as winter closed in on them. Charles and his forces then made for Glasgow, remaining there for a short time. He found comparatively little support there, among the citizens who had previously espoused the anti-government cause of the Covenanters more than that of the Jacobites. Without too much delay he left Glasgow and headed north for Stirling and Falkirk. At Falkirk he met the English detachment of the Government troops, which the Jacobite forces promptly engaged. The Highlanders put the enemy to hasty flight, but Charles and his forces did not pursue them and the advantage was not driven home.

Meanwhile in London the Government, regaining its composure, was recalling troops from Europe and throughout the country. It had assembled a large and well-trained army and was moving it north to intercept the Highland force. Prince Charles fell back on Inverness, and prepared to meet this British force under the command of the Duke of Cumberland for the decisive encounter. The battle ground was to be Culloden Moor, outside Inverness. The choice of ground was not at all to the liking of Charles's adviser, the realistic Murray. He pointed out that this open flat land was not the kind of ground on which the Highlanders fought to their test advantage. The space between the Highland forces and the enemy, when they would be drawn up in battle, would be too great for the full impact of the Highland charge. But the forces were now hungry, they had marched long, they were tired, time had run out, they could not pick and choose. Circumstances had imposed the site of battleground upon them.

Cumberland began the attack with a long and heavy artillery barrage which cut down the Highland ranks. The Highlanders could make no effective reply. They had no equivalent batteries of gunpower. The Highlanders formed to a pattern as in all their other battles. The line was in three ranks which was the usual formation of attack. Each clan massed itself together. The Chief led this group, his closest blood relations behind. In the rear of this group would be the ordinary clansmen who would be less well-equipped. The line of attack then was formed of small irregular columns, each under its own Chief. The whole design of this fighting pattern was destroyed as the Highland forces stood awaiting the belated command to charge. The grapeshot, the gunfire from the British army facing them scythed through their ranks, splitting them up from each other into groups and forces who were unfamiliar with each other's style of fighting. Eventually the Highlanders received the command to charge. "They came running upon our front line like troops of hungry wolves," reported a British soldier who wrote to his wife the following day. On this occasion the Hanoverians were ready for the Highland charge. They trapped the Highlanders between two lines of Cumberland's men. They were then caught in the crossfire and shot down in hundreds. Those

In the midst of harvest

An old pony still working at harvest

Bonnie Prince Charlie,
the 'Young Chevalier,'
coming down the High Street
of Edinburgh

The Prince in prayer
with his Highlanders
before the rout of the
British army at Prestonpans

The Highland army passing through Scotland in high spirits

The final slaughter at Culloden

Ponies at harvest on a Highland estate

Carrying peat to the croft house

who were not shot down at this point became easy targets for the British mounted dragoons to ride down. The scene was watched by the Prince on a knoll of land. Before him stood those units of his army which had not yet attacked. Around him were the last horse troops which formed his guard: gentlemen of property, supplying their own arms and horses, remnants of the Highland gentry which were such an essential ingredient of the Jacobite forces. Many of these were cut down in the continuous lethal fire. Now there was little more than half of the troop of lifeguards left around Charles. The Prince, mounted on his grey horse, was an obvious target for the British gunners who switched their fire to him and his group. A cannonball struck the ground beside Charles. In an account of the incident afterwards he was quoted as saying, "I was riding to the right wing, my horse began to kick, at which I was much surprised, being very quiet and peaceable formerly, and looking very narrowly to him to see what was the matter with him, I observed blood gushing out of his side. 'Oh,' says I, speaking of the horse, 'if this is the story with you, you have no less reason to be uneasy,' whereupon I was obliged to dismount and take another." He mounted a garron—a grey gelding—and rode off to the right of his forces, entering among the dead and dying of his army. He had not proceeded far when he was ordered back, lest the sight of his standard might induce others to follow and cause further confusion among the remainder of his ranks. The struggle could be borne no longer by the clansmen. Many abandoned all hope of life and made futile rushes on the British gunners, apparently in the full knowledge that they would be shot down or cut to pieces in the merciless slaughter.

The battle, of course, was lost. Prince Charles watched the rout of his army in tears. It was O'Sullivan, one of the "seven" who had accompanied him from France and had celebrated with him when he had first set foot on mainland Scotland at Moidart, who grasped the bridle of Charles's garron and dragged his prince away. With a small mounted escort of Highlanders he rode westward, crossing the river Nairn. The small group were well able to keep on the move. The fast movement of this small mounted force across the Highlands as they pursued a complicated route of escape prevented their detection and capture. The loyalty of the Highlanders was such that no reward was great enough for any one of them to reveal the presence of the Prince among them and to break faith with a cause that now would seem to have terminated com-completely. It was the end of a romantic adventure. It was also the end of an era. It was the beginning of a brutal destruction of the clan system. A crossroads in Scottish history had been passed.

Through all the hazards and hardships which followed Charles's escape from Culloden he never showed any loss of cheerfulness and personal courage. Wherever he went he still continued to win the hearts and devotion of all the people he met. In a sense this was his greatest personal fulfilment. Even in defeat he still received the complete affection and faithful support of the Gaelic people.

Eventually, on the 19th September 1746, after many adventures in the Western Isles and a romantic escape to Skye and the mainland aided by Flora Macdonald, Prince Charles decided to leave Scotland. He was picked up by a French frigate from the shores of the same loch where he had landed only 14 months before. He sailed to France, spending the rest of his life in an unhappy exile. The Jacobite cause was ended.

One wonders still about the extent to which this rebellion, which viewed with cold logic was destined to failure, was based upon the figure that Prince Charles presented to the Scottish people: the young horseman who came down the High Street of Edinburgh amidst the throng of civilians and Highlanders who surrounded him, mounted on a beautiful Highland steed, showing him to the utmost advantage as a handsome, dashing and fearless royal adventurer. He was just such a leader as the Highland people and other Scottish communities could accept as a monarch; the figure of a credible Scottish leader made up not only of charm and aspirations but a personality charged with the rapport he had with his men and the public; the figure of the "Young Chevalier."

Chapter VI

A NATIONAL HORSE DISPERSAL

Within a few weeks, between the raising of the standard at Glenfinnan and the assembly of the Jacobite army at Edinburgh, the Highlanders had formed a mounted force of their own horses numbering about 300. These were mainly garrons. They would be aged between three and eight years old, to be most suitable. They would be required to be trained for the saddle and to be wholly sound in wind and limb. This select population, applying modern knowledge of equine population dynamics, would represent about 25% of the available horse population. The catchment area for the Jacobite horses was the Western Isles and the Central Highlands. Geographically this would represent less than half of the Highlands using the garron. Since horses were being mustered after harvest time, crofting horses would be immediately available. Crofters' garrons, judging from latter day population patterns, would be perhaps half of the Highland ponies. As a result of these estimations the total population of horses in Highland Scotland in 1745 must have been approximately 8000. This horse-power was, of course, the essential backbone of all systems of farming to the north of the Highland line, illustrating the enormously important role of the horse in Caledonia.

Following the rout of the Jacobite army, a punitive campaign of retribution against the Highland population was begun on the orders of the Duke of Cumberland. Deservedly he quickly acquired the nickname of "the butcher." All the wounded Highlandmen from the battlefield and those attempting to escape the district were put to death. The Government forces swooped down on Inverness and many citizens had to flee and hide for their lives. This was only the beginning. The bloody persecution of Highland society was to be pursued over a much longer period of time. It was to take the form of a highly organised campaign, which was never apparently frowned upon by the Government in London. It was a campaign of terrorism which took the Government forces into every remote village and township throughout the Highlands.

In the early days following the battle, any man in Highland attire who resembled a rebel was captured and put to death before questions were asked. Even children and women were in danger of their lives. For a substantial period of time the British army maurauded throughout the Highlands and almost at will murdered and slaughtered groups of rural people. It was some time before this campaign of retribution, out of acute national embarrassment, lost its attraction to the British forces.

The occupation troops were quartered at key points to set up defences against another possible Highland rising. In time the Highland occupation changed in character. Having lost their interest in persecuting peasant peoples, the troops took to other activities such as horse races. They found that the Highland pony made an excellent riding animal and organised racing became established in the vicinity of the greater fortresses such as Fort Augustus. For a time this formed the principal diversion for the British troops. In due course, however, their officers felt that the activity was diverting their attention from military duties, and to a large extent it was suppressed. In its place horse coping became a business. Groups of soldiers searched the countryside acquiring horses, some of which had been running stray since the battle. Other horses were confiscated from crofts and Highland estates, while still more were taken by force from unwary riders. A group of soldiers in one particular regiment went into the horse coping business on a very considerable scale and drew so much attention to their activities that again orders had to be issued to suppress a practice that seriously interfered with their normal duties.

News of these happenings reached horse and cattle dealers in the south of Scotland and in the north of England. Dealers who made their living by grazing and maturing such livestock on their grasslands were immediately attracted. These dealers travelled to the Highlands to purchase, at ridiculously low prices, cattle and horses which had been set-up for auction and were being offered for sale by the army. These were the Highland ponies which had been confiscated from crofting peoples and Highland estates. The animals changed hands numerous times, the cost mounting each time before they reached their ultimate destination, usually in the South of Scotland or in the heart of England. Highland ponies began to appear even as far south as Cambridgeshire and in other parts of England about this time. Drawings and other pictures of this era clearly show that the Highland pony was quickly put to suitable use in the rural communities where they eventually found homes. Their suitability for draughtwork and their general versatility made them a popular type of horse among these rural English people who had not previously encountered the breed.

The destruction of the fabric of social life in the Highlands proceeded. Not only was the carrying of traditional weapons such as small daggers banned, but ordinary Highland garb, tartan clothing, any form of dress that might have been associated with the Jacobite cause was banned. Even the bagpipe was outlawed as an instrument of war. At this distance in time, such laws would appear ridiculous and amusing were it not that they remain tragic reminders of the way Scots, and the Highlanders in particular, were metaphorically emasculated as a race. But for those who survived this destruction of homeland, of kith and kin, there came a slow return of life-style and self-pride. Their independence of mind and spirit survived the indignity of the social surgery they so long endured.

In the course of this campaign of attrition which lasted many years, the Highlands had been denuded of their livestock, their horse population in particular. But, being essentially conservational and guardians of a horse which reflected themselves in so many ways, they set out to re-establish the native pony.

The persecution following Culloden sped the final break-up of the centuries-old clan system and paved the way for the Highland clearances. The Government-backed programme of forced depopulation of the crofting counties lasted many decades. From the late eighteenth century it extended to the dawn of the Victorian era in the mid-nineteenth century. The clearances began in the North East and inexorably progressed to the Hebridean Islands. Entire townships from every region of the Highlands were combed out for grasslands to be made exclusively available to the putative landowners for sheep grazing. It was an unexpected and further cruel twist of events for a people who were fatherless, leaderless and already totally subdued, community remnants clinging tenuously to subsistence living. In the words of a contemporary poet, "no seer foretold the children would be banished, that a degenerate Lord might boast his sheep."

They were transported, which was the euphemistic term for their forced banishment overseas to Colonies. The majority were shipped to Canadian shores settling in Nova Scotia, Newfoundland, New Brunswick and Prince Edward Island. The stranded crofters applied themselves to survival in a land not altogether alien in form. Its winter was colder but its summer warmer. The allotted land at least was more arable. This fate was acceptable. "Fair these broad meads, these hoary woods are grand," wrote their poet, "but we are exiles from our father's land." Some of the later clearances took the form of complete coastal township communities being shepherded with their livestock and crofting implements into sailing ships standing offshore to receive them. Inevitably some breeding ponies were loaded and transported also. The details are not always on record, but the folk stories match on both sides of the Atlantic where they are still told without pride.

The Highland pony in the new continent took the change to the western Atlantic coastland in his stride. This horse worked well also on the more open land. He had suitable size, endurance and gait for working among fenced cattle as a saddle horse. Some of his kind were taken south and west where they made their modest but significant contribution to the formation of the cow pony and its later refinement, the Quarter horse. The latter is today a very highly developed uniform and specialised breed, but a familiarity with the Highland pony allows a close observer to have a strong suspicion that there is a trace of Highland blood in it from an earlier date.

The detachment of Lovat Scouts which left for the Boer War at the end of the nineteenth century took a large number of excellent mounts. Paradoxically, though they were Highland ponies, they had been carefully selected for maximum size. Most were over the recognised limit of height of 14.2 hands for modern but totally arbitrary pony classification. The distinguished service of this force and its mounts could not reflect credit on the home breed, of which they were legitimate members, since they were not seen to be true representatives. Whereas existing height definitions might adequately encompass the other eight breeds of British ponies, they are less than adequate for the Highlander. Size variation is not only considerable in this breed but it is one of its main features, helping to make it a breed of almost limitless versatility. Height suppression in a horse breed is an adaptation to inclement wind velocity, in-breeding through insularity, and chronically low levels of nutrition, and it is sad to see breed society policy being added to such a list of negative factors.

Subsequent, more orderly peacetime exportations, during the century to follow, took Highland pony breeding stock to U.S.A., Australia, New Zealand, South Africa, Iceland, Canada, European countries and most notably to England, where the breed is now very well established.

The mainland ponies, especially those grey in colour, found increasing favour in the south of Scotland at the close of the eighteenth century. They crossed nicely with the draught horses of the arable farms there. Their crosses made good dual purpose animals, for riding and for seasonal work. The Moss-troopers of the Scottish Borders used only such grey horses. They set a model for military horses in the south-east of Scotland, later to become Scotland's premier cavalry regiment—the Scots Greys. In the south-west similar "Bonnie Greys," or dapple greys, were also bred from Highlanders. They were featured in Ayrshire. The famous grey mare Meg, described by Burns in "Tam o' Shanter," was representative of such a type—"a finer never lifted leg." It was a similar grey which Burns's Old Farmer saluted on a "New Year's Morning." Burns certainly knew the Highland pony breed appreciatively. He wrote:

> "I paint thee out a Highland filly,
> A sturdy, stubborn, handsome dapple,
> As sleek's a mouse, as round's an apple."

He knew its romantic past, its style of gait:

> "Thou once was i' the foremost rank,
> A filly buirdly, steeve and swank,
> An' set weel down a shapely shank,
> As e'er tread yird;
> An' could hae flown out owre a stank,
> Like ony bird."

The Caledonian horse, in diminished numbers, could never again participate so intimately in the work of crofting families. Nevertheless, it continued to remain visible in the background of those who lived in the North. Throughout much of the year, and certainly throughout the long winter months, small herds of ponies, each of which would be the property of individual crofters, could be seen moving around the hillsides of their own territory in a Highland district. They did not require to be fenced in for they grazed within limits which they set themselves as their home range. These limits gave them adequate grazing land, availability of good fresh streams of water and corners of shelter in spells of severe weather. These horses now had only a pastoral role within Highland life. They seemed almost to be awaiting a fresh demand upon their services by mankind, by their fellow mortals, the Gaels.

From the mid-nineteenth to the mid-twentieth century there was a longer period of peace in the Highlands than had been experienced for a long time. This peace was due in some measure to the fact that the momentum of social life was running down in general. Industrious crofting was no longer able to continue with the loss of so much of the population. Crofting townships became, in many cases, non-viable agricultural areas. They had lost their essential communal work force. Even when the overseas "transportation" of the Highlanders, as part of the clearances, had come to an end, the industrial revolution in the South drew them from their Highland homes. In Scotland the industrial revolution was centred on Glasgow. Numerous jobs were offered there to those manual workers who chose to commit themselves to industrial tasks, forsaking their crofting knowledge and skills.

The depopulation of the Highlands continued. Although precise figures for the period are not available, it has been estimated that by the start of the twentieth century the Highlands were reduced to about 25% of the population which it had supported a century before. There were numerous consequences of this depopulation. Wheelwrights, for example, became very few in the community. This made the replacement of worn out vehicles difficult, and in some cases impossible. Farriers became few and far between and the unshod horses could not be worked. The common grazing lands, which had been so proficiently grazed to their full potential during the summer periods for the livestock of the community, were not in full use. In due course these common grazings lost their value. Vast areas of otherwise good grazing became lost through overgrowth. To this day expanses of bracken in the Highlands are monumental evidence of former grazing land which supported the livestock of entire townships of crofts. With reduced common grazing the crofting people who remained kept fewer livestock, including working ponies. The scope of crofting was further reduced. No longer could the same effort be made to extract from every square yard of potentially fertile soil some modest crop, which would help to feed the people and the livestock during winter.

The depopulation of ponies was very substantial, but an attempt was made to sustain the breed by the Department of Agriculture for Scotland. By providing good quality fertile stallions to parishes and districts in the Highlands and Islands, pony breeding was still carried out. This was of great assistance in maintaining the breed in viable numbers. Not only did this policy maintain a critical level of equine population, but with the breeding of garrons now being done by stallions licensed by a Government Department there was an assurance that stallions of quality were in general use. At times there were complaints that some of the stallions sent by the Department to travel areas of the Highlands had some Clydesdale features in them. This was not desired by crofters and the animals never became popular. The crofting people still preserved the desired type in their Highland mares, and they had a keen eye for suitable stallions. Clerks of crofting townships, where Highland pony breeding was still actively carried out, were usually under pressure from crofters to demand from the Department the stallion of their choice for service in the following season. Not all of these requests could be met, but in the main the better stallions were allowed to make an adequate contribution in the locality before being moved on. Naturally, they were never allowed to remain in an area longer than three years because of the likelihood that they would be breeding some of their own stock. In-breeding was not approved of by Highland society.

The number of people who were able to castrate colts was very limited and as a result entire colts were frequently sold as yearlings, together with cattle, at the regular community sales. These colts were subsequently taken to mainland and lowland Scotland for further trading. Even ploughing was done with one horse where two horses ploughed the same land previously. The breed could meet this demand, though it required a great effort from them in many cases. But the total amount of ploughing done became less. As a consequence, less corn was grown and there was less livestock feeding in the winter season. This tended to be a vicious circle with many crofters ending it by deciding to keep a horse no more. Were it not for the quite profitable sale of yearlings, it is doubtful if the horse population of the Highlands would have been maintained, even at a modest level. That there was still considerable Highland pony breeding taking place in the Highlands and Islands in the early part of this century is evident from the statistics relating to the breeding work of the stallions of the Department of Agriculture. Knocknagael Marksman, perhaps the most famous Highland stallion of all, sired 500 foals in his lifetime. He worked as a travelling stallion for many years in the Hebrides before taking up residence in the Department's stud near Inverness where he continued to breed until the year of his death when he was 29 years old. Marksman, as a prolific sire of Highland foals, serves to illustrate the effort that was being made during this time to preserve quality in the breed.

With the emergence of the Victorian era the Highlands acquired a new and quite spectacular popularity. For the first time they received enthusiastic patronage from British royalty. Queen Victoria became enamoured of the Highlands and made visits to various parts of the Highlands whenever she could. Eventually, Balmoral became her Highland home, and it still remains the Highland home of the British Royal family. Queen Victoria became familiar with the Highland pony and its merits. She is known to have ridden this animal during her nature trails into regions of the Highlands which had no roads suitable for a carriage. The gentry of England followed her royal example and made a recreational invasion of the Highlands. Grouse shooting on the moors and the stalking of deer in the vast deer forest emerged as new forms of sport for the gentry. Both these sports required the back-up of the Highland pony. Grouse shot in large numbers on the high and remote moorlands required transportation to the shooting lodge. They were carried usually in the wickerwork panniers draped on either side of a garron pony, in the way that peat had been transported across rough Highland ground centuries before. The work of deer stalking was ecologically valuable. This sport, which represented a substantial Highland industry, would not have been possible without Highland ponies. The garron was well suited to transport the sportsmen—who were not necessarily in good physical condition—to the remote areas of the Highland deer forests. They were also able to transport back to the shooting lodge the deer which had been shot in the mountains. These deer were mostly stags, very considerable loads of dead weight. Their transportation over many miles of trackless Highland terrain represented demanding work, one that no other type of pony could have performed.

Deer-stalking is still an industry in the Highlands. The red deer population of Scotland which it supports is conservatively estimated at 150,000. A recent report from Leith docks states that, in one year, over 7,000 carcasses of venison were shipped from that port to Germany. Most deer shooting is in the form of professional stalking and amateur shooting, the garron being used as the source of porterage. The magnitude of the deer stalking sport in Scotland indicates the number of garron ponies which are very actively engaged in draught work. Royal participation in deer-stalking has assisted in its promotion. The sport allowed King George V to become familiar with the attractive qualities of the Highland pony, and it is well known that of all the mounts at his disposal his favourite became a grey garron by the name of Jock. Jock can be seen in many a royal photograph, splendidly groomed with his mane and tail pulled. One can see in this example that the Highland pony can make a good-looking—indeed a most attractive—mount.

Classes of Highland ponies included in agricultural shows throughout Scotland have continued to increase, largely as a result of the interest which the public take in these animals when they are on display. These shows also have helped to keep the breed alive by encouraging enthusiastic amateur breeders to continue to breed Highland ponies of the best possible type. They might achieve for their owners the distinction of winning a class in their own local shows! The principal agricultural show of Scotland is, of course, the Royal Highland. For many years this show was held in a different city each year. This also allowed the urban people in Scotland to view at close quarters the stock that were native to their native land. The vision of Highland ponies became imprinted on the mind's eye of many city Scots who might otherwise never have seen these animals. This feedback of general interest in the breed must surely have encouraged some breeders to remain in business and to continue to produce stock of the highest possible quality.

That the Highland pony today is of a very high and uniform standard of quality is due largely to the earnest efforts of those breeders who stayed in the business of garron breeding throughout the years when there must have been little or no financial profit in the enterprise. By the middle of the present century, the massive conversion of farmwork to tractor power virtually terminated the role of the garron animal as a worker on the croft. The change was rapid and for several post-war years the Highlands and Islands were combed by horse dealers buying up garrons at extremely low prices. Few asked these dealers where the horses were destined to go. Those who sold them had no further use for them. The animals, by care and selection, had long life spans and the crofters now could not withstand the considerable cost of maintaining them for long periods when they would be totally unproductive. They were therefore sold off. That many ended in European countries to ease the post-war famine in meat was without doubt. But had this been known it is questionable if the sale of these animals would have taken place on the scale that it did. The post-war clearance of Highland ponies from crofting communities did not last long, however, for new opportunities to make use of Highland ponies developed. Not only did the showing of livestock become active, but the new sport of pony trekking became popular in 1953. This new recreational activity heralded the return of general interest to unspoiled country where only horses could traffic.

Chapter VII

HIGH DAYS AND HAY DAYS

On the cold sharp spring mornings, when the ground sparkled with a light frost upon it, crofters would rise earlier than usual and check that the weather was dry and suitable for ploughing. Then they would prepare their garrons for a day of hard work ploughing those parts of their crofts which were to be used that season for growing corn. Depending on the nature of the ground to be ploughed and the horses available, a pair might be chosen, or a lone garron would be given the job of drawing the plough. One horse would often be given the job on smaller crofts. The horse would have been stabled over-night to be available for an early start. He would be taken out of his stall, bedding brushed off; his feet would be picked up and the wet bedding and manure, packed into the soles of the hooves overnight, would be scraped out. The animal would be led out of the stall by the forelock. The collar, turned upside down, would be pushed over the face and head of the pony and turned around before being settled down onto his shoulders. The bridle would then be fitted and the strap over the back put in place to carry the traces upon metal hooks on each side. These traces would lead back from the collar to be fixed upon either end of a swingle tree, a wooden cross-piece to which was fixed a solid iron ring at its centre; to this the plough would be linked. Long fine ropes would be stretched as plough reins, from the bit on each side, passed over the hames—the iron yokes secured around the leather collar—then backwards, eventually to be carried in either hand of the ploughman, while he manipulated horse and plough.

When the animal was suitably harnessed and the loose pieces gathered up, the horse would be walked to wherever the plough had been left lying from its previous use. The animal would be backed up close to the plough and harnessed to it. The crofter ploughman would then set out for the day's work, turning the plough on its side so that it might be dragged more easily by the pony to the place where the ploughing was to begin. As close to the edge of the field as possible, near the low dry-stone dyke, the man would line the horse up and urge him forward. The pony, doubtless recalling the whole purpose of the exercise from a previous occasion, would push his weight into the collar, drawing the plough quickly forward. The crofter would lift the handles of the small plough upwards so as to drive the tip of the plough share into the subsoil. In many cases the subsoil of the croft would be no more than a few inches deep. When the plough was down to its adequate depth the pony would be kept going forward at a steady, measured tread.

The first furrow was important. It was the line which the subsequent furrows would follow. The crofter would take a series of sightings ahead of the horse, guiding the animal from one to another and attempting to pursue a straight line. In certain circumstances ploughing on the croft demanded a different tactic. Furrows had to follow the contours of the rises and falls of a small croft field. This was important to prevent furrows acting as gutters for rapidly running water when heavy rains fell and flushing away precious soil. At the far end of the ploughing area the furrow would cease, the crofter, by some skilful manoeuvring, would turn the horse around some few yards away from the initial furrow and set off in the direction he had come to make a second furrow going in the opposite direction. By passing up and down the field, the up furrows lying to one side and the down furrows to the other, segments of the ploughing area would be turned over and in the course of the morning perhaps a quarter of an acre would be ploughed. The work would end for the midday rest with a call and a wave from an aproned figure by the croft house.

Highland ponies make good ploughing animals; they enter into the spirit of the exercise. They do not respond to it unwillingly as though it were some chore, but the kinship with man, the apparent knowledge that they are participating in an important ritual seems to give the animals a style, a purpose, in which they fulfil themselves.

When ploughing, garrons will nod their heads, shaking loose their dense manes, and they will snort with satisfaction as they push into the collar. They lift their feet cleanly and dig their hooves into the earth, giving considerable horse power to the operation. As the day wears on and they become heated with work they will arch their tails, in a way that adds to their style, to cool themselves by this action. Thin films of sweat appear on their bodies and the sides of their necks may become lathered. In this work, as in no other, the Highland pony is able to show his grit. With the crofter following behind his horse giving adjustments in direction, manipulating the plough so as to take the right depth of soil and the right width of furrow, his respect for his animal would grow with the passing of the working day. The accomplishment of man and beast would be substantial in the work of ploughing an acreage of corn which, from previous calculations, could provide grain for man and beast throughout the winter after harvest.

At the end of the working day the plough would be uncoupled by the side of the field, the chains would be looped up to the harness, the long reins coiled and thrown over the hames, and the animal would be led back towards the stable. The walk back would be slow, allowing the animal to cool out. Reaching the stable the harness would be removed, the crofter would stop and snatching some old dry pieces of bracken or some hay from the small stack yard by the stable, he would rub the animal down to dry him off. The brisk drying off would be the opportunity for communicating to the horse the crofter's affection for his animal. Soft Gaelic phrases would accompany this grooming. With the animal dried and cooled the crofter would lead his pony by the forelock to some brisk-running stream of mountain water where the animal would slake its thirst. The pony would then be led back, probably to a grassy area within sight of the windows of the croft house. With a noose of a long tether slipped around the leg just above the hoof, he would be allowed to graze for a spell. At his own table, and while eating his own meal, the crofter would keep an eye on his pony grazing around the restricted circle of the tethered area. In a while the crofter would return to the animal, remove the noose from the foot and lead the pony back to the stall, putting down deep fresh bedding and throwing a sheaf of corn into its manger to give sufficient food for the night. The great day's work was ended and the new year of work for the horse, following the long winter-time holiday on the hillsides, was begun.

Days in the hayfield, in the best of summer, were long and filled with back-bending work with sickle and scythe, rake and fork. After the haymaking had gone through its stages of cutting, turning and raking, haycocks in increasing size were built. Horses then joined the families of haymakers in the fields for the final work of loading the hay onto carts and gathering it in, close to the byre.

This was the best time of the season, a happy time when the hay was finally dried and could be securely stacked to maintain the livestock for next year.

The garron has a natural aptitude for deft cartwork. On the field he could be manoeuvred easily by man, woman child or from one haycock to another as the load was picked up. With well fitting harness, such as padded collar and saddle, the pony worked easily and willingly between the shafts of a hay cart, or a boxcart, when the animal could use his power. Negotiating a wheeled load by a difficult route was a test of manoeuvrability that was all in the day's work for a sure-footed garron.

At the end of the day the pony would be led to the spot where the cart was always put up. He would be uncoupled from the shafts and these would be raised to let the horse out to be stabled and unharnessed. After this the several needs of the horse would receive the crofter's attention. The working horse needed watering, feeding, grooming, bedding and securing for the night. Before going to his own bed, the crofter would revisit the stable or trevis and while giving a further supply of feed, would take a few moments to run a hand over his animal appreciatively giving him a hearty pat on the rump for the day's effort. From a normally undemonstrative keeper this gesture could be an adequate reward for a knowing garron, helping in another night of constraint. Later giving himself up to several hours of lying in the man-made bed, the pony would sleep, napping periodically through the night to be wide awake by dawn. By the arrival of the sunlit morning the pony would be recharged with vitality for another day between shafts, in the service of his people.

Cartwork, for a horse with a suitable build for it, is not the worst lot in life. It gives a horse an outlet for its basic urge to move, to keep its limbs in action and to generate one horsepower of energy. Carting comes as close as any other horse-game to the provision for a natural need of dynamic expression. When they are working within themselves horses of the right disposition can derive evident satisfaction drawing carts. The box cart had short shafts so that the animal was well coupled to its load. The box was very close to its hindquarters, sometimes sloping slightly over them to make the draught work a balanced mixture of pulling and carrying. The snug fit between the shafts was created with good

The pony at work shepherding

The rich culture of Highland life surrounding the garron

Peat-carrying on the Isle of Barra in the Victorian age: only the old and very young were likely to ride garrons at this work

Bringing in dead stags in the Edwardian era

The pony in early deer-stalking days

Ponies at a local show on the Isle of Mull at the turn of the century

Photo by courtesy of Scottish Museum of Antiquities

Lovat Scouts watering their large garrons at Nairn after the Boer War

Lovat Scouts at camp in Inverness-shire about 1908

harnessing. The upper side of each shaft carried an iron bar with a hook or ring at either end and a sliding fixture between. The front hook was connected to the harness on the collar, the back hook to the chains of the breeching, the strappings around the hindquarters. The sliding hooks carried the chain which ran over a metal groove on the saddle. The harnessing was completed with a belly-band, usually a soft rope, which passed from one shaft under the horse's chest to the other shaft, to hang loosely and ensure that a movement of the load backward would not cause the shafts to be lifted and the cart tipped. All of this arrangement is, of course, the basic harnessing necessary in securing a horse to a wheeled vehicle. In coach and carriage work the leather and metal harnessing is of quality finish, appropriate to the circumstances of use.

Snugly harnessed in a boxcart, the Highland pony's long tail was a disadvantage. It could not be properly lifted when evacuating and so became soiled. It would get out of the breeching occasionally and could not be drawn in again by the animal, which found this irritating in time. A lengthy tail could get in the way when the box was being released from the frame so as to be tipped up. When the emptied box was dropped back down again, a long tail could become painfully crushed. So the tail was cut or "docked" to a handspan in length when carting was the animal's destined lot. It could be a self-fulfilling destiny for, once docked, the horse could not easily keep off the flies in the summer or retain the warmth between its hindquarters in winter. Docked horses needed stabling and got the use that went with being available. Full-time cart work ideally called for gelding temperament and some garrons possibly found their carting much sweet sorrow.

In the work of driving cattle from the Highlands to the Lowlands of Scotland and the Midlands of England, Highland ponies had the privilege of being companions to the droving men. Drovers plied their trade, moving cattle on the hoof, from the Highland areas of cattle production to the urban centres of consumption. Most of the drovers had their own routes. They would begin their drive from places as distant as Ardnamurchan, Kyle of Lochalsh and Dingwall. Their journeys would end by Glasgow, Edinburgh or pass further south into England ultimately to trade the cattle with butchers and graziers.

Much of the preparation for the drove would involve the pony.

Over a saddle would be placed a small, well-padded wooden tressle. Upon this would be fixed the numerous items essential to the drover in his long journey. A bag of coarse oatmeal would be a necessity, not only to feed the drover himself, but also to supply the pony with extra nourishment from time to time, especially on the return journey. When the drover had his business finished he would ride much of the way home, without too much time to graze the pony while trying to beat the winter.

Some essential cooking items would be added to the pack. More than one blanket would be rolled up and bound onto the pads. At least one plaid would also be included with an extra pair of boots, for much of the southbound journey would be on foot; drovers knew the wisdom of sparing their garrons for as much of the journey as possible. With a heavy stone-ware jug of whisky secured, the whole pack would be draped with a sheepskin. The drover would set off leading his garron on a rope halter with a long shank which allowed the pony to follow behind. With his dog and his stick he would direct the course the cattle would take down the long glens that led to the grasslands of the south. One drover could manage about twenty head. Bigger droves of cattle numbering over a hundred head would be driven by an experienced man with the help of younger men willing to work on foot as directed, eager to enjoy the adventure and learn the trade. Smaller droves were more usual, for drovers preferred to work alone. They also found it easier to get grazing for their animals—as they stopped the drive about the middle of the day and in the evening—when numbers of cattle were limited. They might ride for the last hour of the drive in the early evening astride the pack, and a drove could regularly cover about ten miles per day without the stock losing condition. On the return journey the drover would walk for the morning and ride in the later afternoon, to cover in comfort twenty miles per day and still enjoy some social activities in towns along the high road home.

As he travelled homewards the drover would accumulate packages of snuff, tobacco, tea and perfumes to take back to his own community for trading and for gifts. These commodities could be acquired in market towns. But in fishing villages along the east coast, where the European smuggling trade was active, bargains were better. Their concealment in the horse's pack was a simple matter. Once a light load of such contraband was acquired a drover would make haste homeward.

On the final laps of the return journey the drover would ride his garron more often, as he tired and the horse quickened. The animal was then in the best of physical condition, hardened by the long excursion; he knew from the northern sky ahead and the feel of the sun on his rump when he was going homeward. His pace would quicken as he got among the hills and saw the mountains of his own heath. He would hear racing streams, the plaintive call of curlews, the screaming of gulls, the sounds of homeland. The nearer the journey came to its close the stronger would be the pulling power of home. The smell of peat lands, the taste of Highland herbage, the crystal clear air, which carried smell and sound from afar, the feel of fine rain, they were all so familiar. The horse's ears would not cease moving in every direction and his nostrils would be held wide open to absorb all the signs of his own precious place on earth.

Chapter VIII

GARRON RIDING

In 1725 General Wade was appointed by the Government in London as the Commander-in-Chief of Scotland, his principal function in the following decade being to penetrate the Highland regions with a system of roads. These roads linked together at strategic points such as Fort William, Fort Augustus, Fort George and Inverness. His road works still remain; General Wade's Military Road runs southwest from Inverness to the south of the loch, and General Wade's Bridge is a beautiful stone structure situated on the outskirts of Aberfeldy. The road building programme was very successful in improving communications throughout many areas of the Highlands which had previously been impenetrable to wheeled vehicles. Apart from the roads constructed as part of the campaign, the policy of an improved road system of communication between Highland townships became adopted, and road building and road improvements throughout the Highlands and Hebrides were implemented. The Highland pony was the principal means of communication throughout the Highlands and Islands with or without roads. The use of the pony was important also in droving. The type of pony required to perform this type of service was not necessarily the same as that sought for draught work around farms and crofts. Horsemen of that era had as much knowledge of the types of horses best suited for riding as generally exists today and the breeding of animals of this type was a notable activity.

Gervase Markham wrote his *Complete Horseman* in 1614 and one of the principal sections of that book is titled "Of Horses for Travel." In this section Markham describes in some detail the points of conformation in a good travelling horse. He states the following: "The marks whereby we shall choose a good travelling horse are these: he shall be of a good colour and shape, lean headed and round foreheaded, a full eye, open nostril, wide jawed, loose throated, deep necked, thin crested, broad breasted, flat chinned, out ribbed, clean limbed, short jointed, strong hoofed, well metalled, neither fiery nor craving, strong in every member and easy to mount and get up upon." In reading this one would think that Markham had in mind the riding type of Highland pony. He does not refer to the ideal height of the animal and indeed this omission is highly significant. Markham's recommendations are as sound principles today as they were then, and they can be as readily applied in the selection of a riding pony of Highland breeding.

In the book entitled *Highland Ponies with Reminiscences of Highland Men* by John M. McDonald, published in 1937, the writer gives his well considered description of good points of desirable conformation and characteristics in this breed. Among the points he lists are the following: "sturdy, somewhat rounded, muscular and powerful animal, about 14.1 hands high . . . with preferably a sloping shoulder well ribbed up, not too long, strong arms and thighs etc." Clearly this general description would apply to most Highland ponies, particularly the mainland type, even today. More particular details of the ideal conformation are quoted as follows: "The neck strong and arched, but not too thick near the jowl, a full mane and tail and what may be called a perfect head; ears short, set up and cocked like an Arab's; bright eyes with the same characteristics and open nostrils; wide in appearance between the eyes, probably rather dipped above the nose, a short sensitive muzzle and, above all, the air of intelligence. In action the pony should be a free walker and when trotting should bend the knee and cover the ground neither too high nor too low." This description, whilst matching Markham's in many ways, is also one which would be useful in the selection of a pony for rough and general riding.

Later in his interesting book John M. McDonald, a native of Skye, describes how as a young man he embarked on extensive travels throughout the Highlands and Islands on horseback. One can see that he was somewhat ahead of his time in recognising the suitability of the Highland pony for nature trailing and trekking in the Highlands. His glamorous excursions are on record and are invaluable in bringing to our attention some inherent features of his ponies. One of his most interesting accounts is illustrative of the swimming ability of this breed. He describes how, in taking a short cut to catch a ferry, he engaged a boatman to row him across Loch Broom. His mount, a grey Highland mare, was led on the loose end of the halter from the stern of the boat, swimming behind it effortlessly across the substantial width of Loch Broom. At the end of the swim McDonald mounted his pony and continued briskly on his way, meeting the ferry in good time. McDonald describes his grey mare as follows: "She was about 14.2 hands in height, had a small Arab-looking head, fine long neck and slanting riding shoulders and her eyes showed that bold yet docile and intelligent look which characterises her breed." As to her lineage, it seems to resemble the uncharted breeding of many Highland ponies. He records it as "the first of the strain owned by my grandfather in the Outer Isles and my father always maintained that she was descended from 'Boisdale's Ponies,' the name usually applied to the descendants of the Spanish horses brought to Uist by Clanranald about 1712." McDonald at another point in his book describes how he preferred his mare hogg-maned, though this is certainly frowned upon in the case of ponies which are to be put in the show ring and is also not to be advised for outwintered Highland ponies. Nevertheless, George V's favoured Highland pony, a grey garron called "Jock," was groomed as a riding animal with forelock, mane and tail well-pulled. It is surprising how this removal of hair can alter the appearance of the Highland pony from that of a working draught animal to one of riding style. It is also interesting to note that both Markham and McDonald emphasised a good slope in the shoulders of an animal which will make a good steed for riding over distances. Perhaps this is one point of conformation where there may be a substantial difference between pony types because one does not get this slope of shoulder in a horse which has conformation for draught work. Another point of difference in conformation to be sought in the animal more suited for riding, is prominent withers. These points apart, one finds that the general conformation which characterises the Highland breed makes it singularly utilitarian.

Some latter-day critics of the Highland pony have made the statement, one suspects somewhat sarcastically, that the best gait of the Highland pony is the walk, but this is not only very true but is a positive asset in any breed for endurance trail work. Other uncomplimentary statements about the gaits of this horse have often been made in ignorance.

In studying the gaits of the horse most authorities indicate the division between the natural gaits and the artificial gaits. Many references, of a definitive nature, exist in relation to this topic. One which gives a concise and accurate account of these is contained in

Ecole de Cavalerie. Chapter 5 in this classical work describes the gaits of the horse. A translation states the following: "Natural gaits are divided into the perfect gaits which are the walk, trot and canter; and the imperfect or defective gaits which are bass, single-foot and rack. Perfect gaits come wholly from nature and are without the elaboration of art." It is relevant, therefore, to focus some attention on these perfect natural gaits, which are described further in the translation as follows, *e.g.* "the walk."

> "The walk has the least elevated, slowest and gentlest action of all the gaits. In this movement, the horse lifts the two legs that are diagonally opposed—a fore leg and a hind leg. When the right foreleg is off the ground and moving forward, the left hind leg is raised and follows the movement of the foreleg. The other pair of diagonally-opposed legs move in the same manner. There are four movements in the walk—first, that of the right foreleg followed by the left hind leg to make the second movement. The third movement is that of the left foreleg, followed by the right hind leg, and so on, alternating between each pair of diagonals."

This description whilst correct is somewhat misleading in that mention is not made of the fact that the walk is often commenced by propulsion of one of the hind legs. The walk is a four-beat movement and the four beats should be evenly separated from each other so that one four-beat movement blends in with the other, each individual beat having equal time and emphasis. As with all forms of gait, the action of each limb can be divided into two parts. These are the "swing time" and the "standing time." The latter relates to the duration of time when the foot is on the ground, while the swing time is the remainder of time when the foot is in flight. During the swing time the feet should be lifted smartly and cleanly from the ground, with good flexion of the knee and the fetlock in the foreleg, and the hock and the fetlock in the hind leg. The foot should be carried forward clear of the ground without the action being exaggerated. Animals which do not have a good swing action are the ones which more often stumble. A limb which moves in this fashion is set down on the ground firmly and does not therefore give an appearance of shuffling. Shuffling and "ambling" are thought by some to be the same type of action, but the amble is a quite specific type of walk and it is a desirable one. Shuffling on the other hand is a defective form of walk. The amble, which, for the rider, is a comfortable gait, is essentially an accelerated walk. In the amble, the increase in walking speed is achieved by the horse lifting, before replacing, the lateral pairs. That is, the following foot is raised just before the preceding foot is placed on the ground. This is obviously a gait which is more demanding than the normal walk and there are few breeds of horses which can match the Highland pony for endurance in maintaining the amble for long periods of time over considerable distances.

In the translation of *Ecole de Cavalierie* the trot is described as follows:

> "The horse trots by lifting two diagonally opposite legs at the same time. Thus, his right foreleg moves with the left hind leg. The difference between the walk and the trot is that the trot is more energetic, active and elevated. This is what makes it rougher than the walk which is slow and close to the ground. There is another difference between these two gaits. At the walk, the horse's legs move on opposite diagonals as they do at the trot but it is a four beat movement. There are only two beats at the trot, because the horse raises and sets down two feet at the same time."

The two-tempo beat of the trot is synchronised with forward movement of diagonal legs. The rhythm remains the same. At the trot the rider has the option of remaining in the saddle and performing the sitting trot. The sitting trot requires a good deal of action and co-ordination on the part of the rider, particularly in the lower back region. Some horses which strike the ground firmly at the trot can be uncomfortable rides at the sitting trot. The more comfortable way of riding the horse at the trot is by using the rising trot, or posting. In posting, the rider rises from the saddle, taking the weight in the stirrups when one or other of the forelegs is placing full weight on the ground. Over a distance, a rider in the trot is advised to change the timing of the rise to alternate forelegs.

Returning again to the translation of the French classic, the canter is described as the movement of a running horse:

> "It is a kind of forward leap, because just when the forelegs are not quite back on the ground, and the hind legs are still

Photo by courtesy of Scottish Museum of Antiquities

A deer-stalking party at the end of a long day in the Perthshire hills

Drover's departure *(Landseer)*

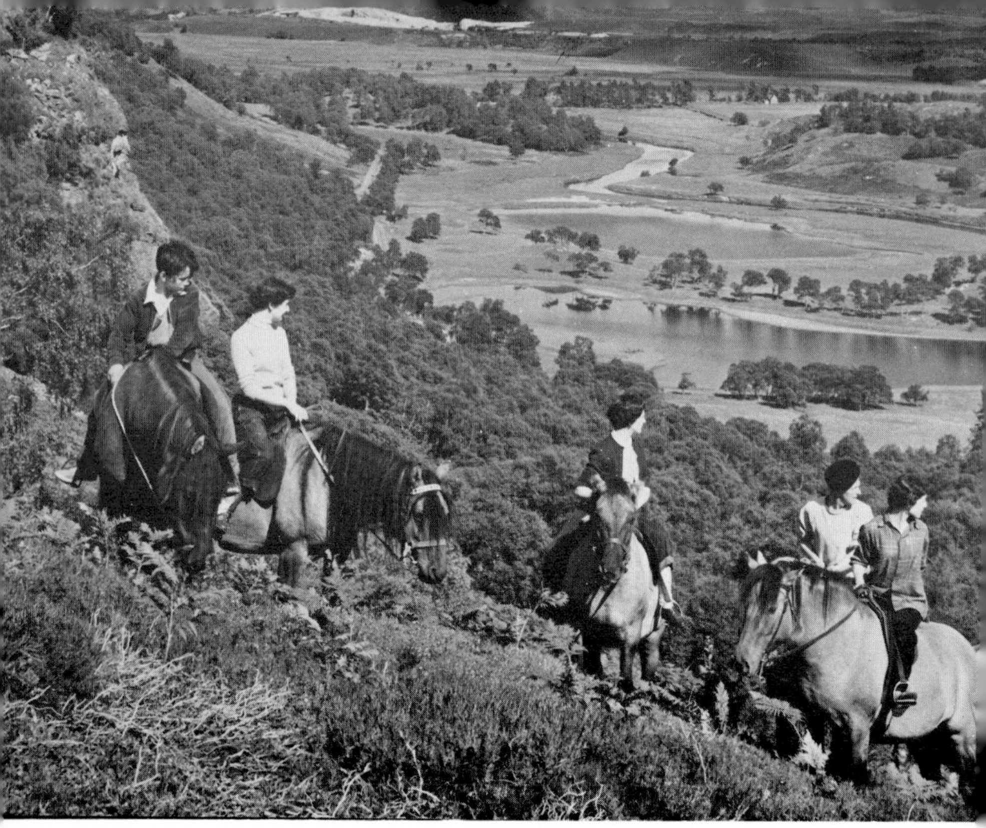

Pony trekkers enjoying the view at Newtonmore

A Highland pony and carriage on a road from the Isles—hopefully not to the oblivion threatening his kind

raised there is an almost imperceptible movement when all four legs are in the air. The canter, which keeps waltz time, is a pleasing gait for the rider as much as it is to the observer."

While the canter is not the garron's favourite gait, being against his principles of economical locomotion, he can move out in this gait in very respectable fashion and pace. Cantering, however, is not the best way to ride in the "land of brown heath and shaggy wood, land of the mountain and the flood," as Scott described and hailed Caledonia.

For a horse to be ridden at a fast gait over an extended distance it must maintain a good system of breathing since its rate of respiration increases with increase of gait. The release of energy within the animal body, in order to give it power to move musculature, is done with the production of heat within the body. Excessive heat accumulated by the animal readily leads to hyperthermia, which, in the working horse, is a serious physiological state if it cannot be relieved. Relief of hyperthermia in the riding horse can only take place in two ways. One is to increase the output of heat from the body through increased expiration of its own warm air, which calls for respiratory efficiency in these features of conformation that have already been described—such features as a good clean throat and a wide muzzle. The latter is required to widen the nasal openings by the movement of the cartilagenous wings of the nostrils so as to make the nasal aperture as round and as open as possible.

The main thermo-regulatory mechanism available to the pony which is becoming warm as a result of exercise is to sweat. Sweating facilitates radiation of heat over wide areas of the surface of the body. A degree of air movement over the surface of the body, as a result of the animal's own progression, helps the loss of heat from the body as a whole. Sweating takes place over most of the horse's body when it is well heated from exercise. One of the areas of the body which can lose heat very quickly and proficiently, when this is necessary, is the region beneath the tail, in the region of the dock and between the thighs. These are the same areas which require to be covered by the tail when the animal is conserving its body heat in cold weather. By the same token this area can act as a thermal vent when the animal is overheated. It is therefore a key part of the thermo-regulatory system, together with conformation, breathing and sweating. When the animal becomes warm as a result of exercise it promptly elevates the tail to allow the dock area to function as a thermal vent, a "flue," For this reason horses suitably warmed up by brisk exercise will raise and carry their tails out from the hindquarters. This appearance of the animal's carriage is to improve dissemination of increased body heat.

It will be apparent that the Highlander, when being ridden over his native terrain, coping with uphill climbs and the additional work of accommodating to uneven ground, needs the opportunity to breathe freely. Under these conditions the rider should dismount whenever possible at the start of an uphill climb and walk with the horse. If the rider chooses to remain on horseback uphill, the horse should be rested after the climb has been completed. The view can then be consumed appreciatively by the dismounted rider. By observing the respiration of the animal it soon becomes evident when it has adequately ventilated itself after the extra effort and is ready to proceed.

When the Highland pony is used regularly as a riding animal it requires no less care in grooming than would be expected in any other breed. The first grooming requirement is attention to feet and this cannot be done too frequently. It should certainly be carried out daily and preferably before, after, and even during a ride. Picking out the feet so as to clean the frog and sole and remove any foreign matter which has become lodged there, is most essential to soundness.

A pony which has broken sweat at some time will have the sweat dried out, and this dried sweat will be retained within its coat. Accumulation of sweat leads to discomfort in the animal and, in due course, to inflammation of parts of the skin. In order to remove all traces of dried sweat, the face of the pony should be properly wiped with a damp clean cloth, using hygienic precautions in doing so. The area under the tail and all of the dock region, the posterior and inner thighs, also require to be wiped and cleaned, again using a separate moistened cloth or sponge in a hygienic procedure. The back requires special attention in grooming, and the use of a body brush over this region is essential. Limbs also require to be brushed and if necessary wiped with a damp sponge when the animal is being groomed. It is important to remove all traces of dirt, dried sweat, etc.

When ponies are not in active use and are running free at pasture they are able to perform much of their own grooming requirements. The individual horse is able to do much of this by itself when it has the companionship of another and they are able to become involved in mutual grooming to their joint advantage. In self-grooming the pony rubs its face vigorously up and down one leg towards the inside of the knee. Again it can groom itself by turning its head round to its back and nipping its skin over its barrel which is accessible. As a form of natural self-grooming activity, ponies will rub their manes, other parts of the body and including their hindquarters against solid objects such as trees, gates and buildings. When the horse has the company of its own kind, mutual grooming between pairs is often shown when they overlap their heads and necks and nip each other over the withers and back simultaneously, and there is evident satisfaction for the pair in this form of behaviour.

In maintaining a Highland pony in condition for riding purposes it is desirable to give it exercise on idle days. Supplementary exercise is best done by lunging. For this reason ridden ponies should be trained to lunge easily. Lunging sessions can be stressful episodes if they are not carried out with due care and consideration by the person lunging the animal. The animal must be taught with care how to respond to lunging. When horses are lunged they should be given equal time in both directions so that they do not acquire a habit of favouring one lead over another or develop muscularly more on one side than the other.

Ponies which are being ridden regularly require more feeding than usual. In addition to good grazing, a supply of concentrated feed must be provided. No pony should be given work, including riding, without additional protein and energy-supplying foodstuffs as fuel. The pony has a small stomach and concentrate feeds should be fed in small quantities, ideally two per day. Of the two feeds, one can be a smaller feed given before the exercise, the other one, the larger part, to be given after. The concentrate feeding can take the form of about 4-6 lbs of bruised oats per day to which should be added 1½ lbs of bran. Whilst grass is growing, from approximately Easter time until the end of September in the Highlands, grazing will provide sufficient nourishment if they have adequate space in which to move about and select their feeding, and if the quality of grazing is satisfactory. At the other seasons of the year, especially during winter and spring, there is a need to supply extra energy in the diet of the pony so that he maintains adequate physical condition. Many pony keepers find that the most demanding season in the year for the pony is early spring, before the grass starts to grow. Although the weather may be less cold at this time, rain and high winds can chill the animal. These two factors, moisture and air speed, have a greater chilling capacity on the animal than merely a drop in temperature would have by itself.

When ponies are on particularly rich pasture they may become overweight, even obese. Whilst obesity in breeding stock seems to find some favour in the show ring, this is not to be taken as a reliable guide of the best condition in which garrons should be maintained. If there is evidence of overweight in ponies during the summer months, they should be taken off grass for at least part of the day, to be stabled or possibly moved into a smaller paddock where there is no grazing. In winter, ponies need hay probably from the end of October through to Easter. Hay should be fed to ponies on a daily basis. In a really hard winter as much as 14 lbs of hay per day should be given to each pony. This is in addition to one concentrate feed per day. In winter the concentrate feed to be given is probably best in the form of compounded rations made up commercially. These rations have additional ingredients such as trace elements and minerals which are not necessarily available in hay or in oats. Compounded rations may not be needed at other times of the year when ponies are grazing on growing grass, since they should acquire these nutrient materials on satisfactory grazings. "Pony nuts" made up to a guaranteed ration, with instructions provided as to the manner of feeding, are by far the quickest way to ensure good nutrition throughout the winter months. If the pony finishes the winter and the spring in poor condition it takes some time, well into the summer period, to recoup the loss of condition. If animals are to be entered in shows—and a great many shows are dated in the early part of the summer—they do not have the same prospect in the show ring as if they had been well maintained throughout the winter period. Other suitable food for Highland ponies includes flaked barley, bran, turnips and, even in some circumstances, raw potatoes.

Pregnant mares also require to be fed concentrated feed in the same rations as would be provided for a working riding pony. When feeding ponies at pasture while they are running in small bands, it is preferable to distribute feed, such as hay, in different locations so that each pony is able to go to a supply quite separate from any of the others. This avoids poaching, bullying and the intimidation of ponies that are low in the "peck order." Ponies that are not well established in the social hierarchy within a herd can suffer from poor nutrition as a result of bullying and deprivation.

The Highland as a breed is susceptible to souring, as a result of boring or unfortunate experiences, and these experiences should therefore be avoided. They should be particularly avoided at the start of a training programme for the young horse. The secret of good training is in the possession of a practical knowledge of horse behaviour. This gives the trainer an ability to understand a horse. The training of a pony is required to take place in a step-by-step progression. The first step is to evaluate the horse's temperament. To do this, the trainer should get to know the horse by studying it and watching it under various circumstances. If the trainer has a poor opinion of the horse he should not proceed further. The first impression of a horse should indicate to its potential trainer whether the horse's temperament and personality is one that is appreciated. If appreciation develops, there is an excellent basis for the stages of training to be undertaken successfully. Throughout the following stages of training the pony's particular mental and physical limitations should be recognised. These will vary from one pony to another, but these limitations are very real. A good trainer teaches the pony to perform well within its physical and mental potential, but never attempts to force the pony beyond its limitations. In particular, confrontations between pony and trainer should be avoided at all costs in the early stages of training. If a good trainer observes that a horse does not respond enthusiastically, or well, to given routine, that routine is avoided promptly, rather than create a bad experience for the animal. The good trainer does use discipline as well as rewards in training but the way that this is done is in a rational manner and at the proper time. Rewards such as vocal praise and reassuring patting with the hand in the vicinity of the shoulder should be done immediately after a lesson and when a new response is successfully learned. It is important that the reward be given as close in time as possible to the successful achievement. Similarly, punishment, if it is given at all, should relate immediately to the reaction of the horse which is unwanted, but punishment should be reserved only for those occasions when the behaviour is highly undesirable or when the trainer knows that a horse has disobeyed or ignored a lesson which was formerly satisfactorily taught and learned. This time sequence between event and reward or punishment is the way in which horses learn best those habits which make them most useful. It is the garron's basic desire to be useful and its education should respect this most praiseworthy native trait.

CHAPTER IX

THE FUTURE PONY

Today as we search for sources of energy alternative to those on limited supply in the world, attention must inevitably turn again to the horse. The Highland pony is one of the most suitable breeds of horse to supply energy in a form of natural resource which need never become exhausted. As an energy resource the versatile Highland pony only waits to be rediscovered and re-used. One such function is as a chore horse on farms, providing power for vehicles moving loads of materials in and around farm buildings. This does not displace the use of tractor power on the farm but it does limit tractor use to the major agricultural role for which it is necessary. The chore horse is a new type of role which is becoming increasingly popular on Canadian farms, for instance. One sees every prospect of the Highland pony filling such a role admirably on Scottish farms and indeed on farms throughout Britain. It would also be an animal suitable for performing routine duties of delivery of materials on estates. The use of the pony in such a way on national agricultural estates in particular would not only be economical but would perhaps set an example of how this animal can be used in a manner appropriate to its abilities and our contemporary exigencies. Of course, the animal has a general purpose as a form of transport: he can be driven or mounted equally well and when used within the confines of a farm estate could be put to considerable and varied use without impeding or being impeded by motor transport.

When considering the question of conservation of the Highland pony, one must realise that since this is the heavyweight among ponies it is the one breed which would be most susceptible to large-scale butchering if its value in the form of edible flesh became greater than its value in any other role. History has taught us this lesson. For this reason there needs to be a system of controls which would safeguard the breed from being depleted to the point of elimination should there be perhaps a sudden and short-term increase in the value of horse flesh which afterwards would be irreplaceable. This to some extent occurred after World War II, particularly among the Western Isles ponies.

In the meantime the Highland pony has entered the Common Market. In some of the EEC countries horse-flesh is a socially-acceptable dish, even a preferred one. Access of such countries to the pony of the Highlands is now more feasible than before. This has to be recognised and some form of protection should be afforded this now vulnerable animal. For example, it would be an elementary procedure to place a ban on the exportation of garrons, allowing only the exportation of highly-priced breeding stock. Furthermore, it would be an elementary precaution to declare that all "eel-striped horses" should be protected against indiscriminate commercial slaughter. If the horse-slaughtering industry set its sights on this pony, which undoubtedly must have good carcass value, large-scale butchering of the animal could occur.

The Highlander as a breed, in its present number, with so many satisfactory specimens to be found, says much for the Highland Pony Society and its members who have striven to retain a good range of specimens and to promote the breed. The breed has not yet won the wide favour that it deserves among the equine-conscious public but that it will do so in time is without doubt.

A case can be made to recommend that a National Pony Council for Scotland be established. Such a Council, supported by Government backing, the Highland Pony Society and Shetland Pony Society, together with representatives of the Department of Agriculture for Scotland, would be a powerful force to protect the interests of all ponies in Scotland.

Another national development which would appear to be worthy of early and serious consideration is the formation of a National Stud for Highland Pony stock. The Department of Agriculture and Fisheries for Scotland maintained until 1978 a small stud of Highland ponies at Beechwood in Inverness. The stallions there were all wintered out and only housed during the service season. Beechwood stud was founded in 1913 by the purchase of two mares, May Mist and Miss Dew. In the course of time the stud was built up by the detention of the best progeny from the two mares. Many of their sons were kept to meet the requirements of Highland pony breeding, particularly in the Western Isles. For this latter work some 18 stallions at one time were necessary to meet the breeding requirements of the crofters. Grooms were hired, suitably equipped and were sent with stallions to various breeding locations throughout the Western Isles. Many of these stallions were prolific breeders. It was regrettable that one has to report that the bulk of the progeny of these great sires have been lost in the post-war years.

Until its closure, the stud stallions at Beechwood were excellent specimens as indeed were most of their predecessors, some of whom represented the most famous Highland ponies in history—animals such as MacPherson, Fender Laddie, Boy David and Knocknagael Marksman. Marksman in fact was considered to have the best head ever seen on a Highland stallion and to this day a mounted specimen of this animal's head is to be seen in the hall of the Royal Dick School of Veterinary Studies in Edinburgh.*

The National Stud for thoroughbreds at Newmarket is a very expensive enterprise, but there is no reason why a similar, if less ambitious facility, should not be created for the principal indigenous breed in Scotland, to provide for its protection and the maintenance of its qualities.

In ensuring that the best qualities are maintained within the breed, stallion selection should be maintained at least at the present level. The licensing of stallions could perhaps be improved by joint examination of proposed colts being carried out by a livestock officer and a veterinary surgeon both appointed by the Department of Agriculture while being assured that their knowledge of the breed was satisfactory.

The recent increase in equitation generally among the British public is phenomenal. Perhaps television has helped in this but at the same time it has to be borne in mind that the public generally is becoming much more appreciative of diversity in recreation. In particular, outdoor and highly mobile recreation is appreciated. The horse allows this to be a sport which is consistent with natural economics and with man's evident need for harmless and even beneficial use of animals in sport. The sport of kings on the race-track has been with us for some time and has represented the main form of interest in equine recreation. Today, however, we are seeing the rapid emergence of a newer sport, which could be called the sport of princes and commoners—that of carriage-driving using

* As a personal note, I am proud to say that I aquired his grandson, the grey stallion from Argyllshire, now breeding in Skye.

teams of horses. It would appear that this new sport had its origin in Hungary. It certainly provides an outdoor spectacle for large numbers of spectators who, without becoming actively involved, need not on the other hand remain spectators. Carriage-driving over lengthy courses allows a group of people to travel out into rural settings to observe teams performing in a sport which is likely to interest an entire family unit in a day's outing. In any event the sport for those actively involved seems likely to be increasingly popular and a Highland pony can play a very considerable role in this. In a recent major driving event I was able to see numerous, very well matched teams participate; breeds ranging from Shetlands to Cleveland Bays and Exmoors. Many others were on show but no team of Highland ponies was represented. It would appear that this breed may be particularly suitable for forming a team which ideally should be of similar appearance, including size. Since it has so much consistency in these features, the Highland breed should be able to provide matched teams with minimal economy. Again this animal, being renowned for its power, would surely be competent to put up a good athletic performance. Furthermore, since this might be considered an expensive sport, Highland ponies would probably be the most economical animals to maintain for this purpose, and it seems highly likely that the breed still has to make its mark in this activity.

Endurance riding is yet another form of equestrian sport which is of increasing interest. In this activity very substantial distances have to be covered by horse and rider within a given period of time. Twenty-five, fifty or a hundred miles are distances that are frequently set. Comparatively few horses are capable of completing a 100-mile endurance ride. Veterinary examination is performed at checkpoints along the way in such rides. In the event that a horse is found likely to overtax itself at any stage it is then pulled out of the competition. At the moment the Arab breed is the one which seems to be generally successful in this type of event which by the way is a test of athletic competence on the part of the rider as well. It would appear that the Highland pony again has still to make his mark here. It would almost seem to be tailormade for such an activity, requiring power and endurance and an economical carriage or way-of-going. At least one endurance ride in the United States in recent years was won by a horse of Highland pony blood.

In general recreation the Highland pony has already established himself. The sport of trekking is now very well established and is progressively becoming popular. An early promoter was Hugh MacGregor, following his involvement in 1953 in a film which utilised a substantial number of riding horses. Today his centre is a very successful one in pony trekking activities. These take place in the Trossachs around Aberfoyle, providing enormous pleasure and recreational opportunity to persons of all ages and abilities. In addition to Mr MacGregor's enterprise, other pony trekking centres have become established, notably in the Aviemore area. The sport of pony trekking is clearly a permanently-established one and is still growing. Perhaps it does not need to be promoted further within the context of this book. One feels that there may be further scope for expanding this type of equine activity if the emphasis takes the form of nature trailing. While nature trailing on horseback might appear to be synonymous with pony trekking, it might be more opportune if greater emphasis is given for the participants to observe the features of nature.

This form of sport, in which large numbers of people ride out on Highland ponies for a day's activity, taking the animals at their natural walking gait, allows group adventuring. Outward Bound activities are yet another development awaiting the use of garrons. Outward Bound Schools are now recognised as being extremely beneficial to young city dwellers, to people who require a broader interest in life. It would seem too obvious that such activities for Highland ponies could be promoted with great benefit to many. One of the great values of the Highlander is that it makes not only an excellent mount for experienced horse lovers, but also it is an animal that can be used by inexperienced riders. It can be ridden with great safety by both young and old. Undoubtedly supervision and perhaps some instruction is necessary, but no sport is without these requirements. From these examples we can see that the role of the Highland pony in general outdoor recreation is likely to increase.

The disabled section of the community deserves every chance to participate in outdoor recreation. The phenomenal improvements in the quality of handicapped living, which have resulted from equitation facilities for the disabled, are now recognised. This has resulted in determined attempts to provide these therapeutic facilities of equitation to physically and mentally handicapped individuals. These schemes deserve the highest praise, for they face monumental difficulties of organisation and financial foundation. The limitations of these schemes to date must be seen to be the result of indifference, ignorance or inertia by the public, perhaps all of these. Greater active support for disabled riding must be created somehow. Included with the more evidently physically or mentally handicapped should be the socially disadvantaged. Age barriers should not exist, for both children and older adults appear to respond positively to equine contact. This activity should be properly described as "therapeutic equitation," for that is what it is. Whilst its most effective form is riding, there should be scope for carriage-driven forms for the benefit of disabled people. It will be immediately obvious that the Highland pony, and indeed the Highland environment also, is the ideal material. Residential centres are required for such a development to take place on a satisfactory scale. They are easily justified and urgently needed. The handicapped and the deprived have a right to the provision of facilities for the pursuit of all sporting activities available to the rest of the community.

The therapy of riding results from improved balance, co-ordination, muscular strength, emotional confidence and greater self-esteem from accomplishment. Ability, not disability, becomes emphasised. *It is Ability that Counts* is the title of a book which is a training manual on therapeutic riding written by Lida McCowan and published by Alexander Mackay-Smith in the U.S.A. The title says much. The *Highland Pony News* in the spring edition of 1971 reported on ponies of "The Drum Riding for Disabled Trust" which is associated with the "Riding for the Disabled Association" in Britain. That report indicated that various types of horse are needed in this work but the Highlander and its virtues are found to be fully appreciated. This latter venture is a small but invaluable model onto which a national Treasury-supported scheme should be grafted. Therapeutic riding is now to be included as a permanent feature of future "Special Olympic Games."

Given such prospects of a good future, as we have been able to see, there would appear to be now an additional need for improved education on Highland pony affairs. Again, the showing of the pony is the principal way of putting it "in the shop window" for the population to observe at its best. These interests require to be more closely looked at than they have been until now.

The formation of regional centres for equine instruction, in Scotland at least, could readily be the responsibility of a National Pony Council. Appropriately qualified instructors would be comparatively easy to find and, judging from past experience, it is likely that their services would be gladly made available at virtually no cost.

From time to time the question arises whether horses would be subject to licensed registration in much the same way that some other companion animals such as the dog are required to be. Some slight expansion or modification of a registration scheme and the addition of a comparatively modest fee might make it possible to create an advisory inspection service, perhaps on an annual basis for all pony owners. These inspection services could be provided by local veterinary inspectors in much the same way that inspection is already done in cattle. Advisory services would allow advice to be given if the breeder has certain questions or problems concerning the horses' welfare. Should such a scheme be formed in the future one hopes that crofters' animals would be permitted free registration at the same time as receiving free advisory inspections.

With increased level of showing activity there is a need for a degree of proficiency in judging. Awards could be given not just on the basis of conformation but also for such considerations as pony care and handling.

Those who come along to observe the animals which win ribbons at shows should know that they are observing the style, conformation and general features of the Highland pony as recommended by its governing society.

We owe it to this breed to show it off to its best. In fact we owe it much more than this. There is still time for us to acknowledge Scottish society's debt to this animal, time yet to hail the garron as one of our national heroes, a staunch ally of mankind in this special part of the world where survival is paradoxically a constant struggle and its own rich reward.

Now that the Beechwood Stud at Inverness, so long maintained by the Department of Agriculture for Scotland, is closed down permanently, the sparse distribution of Highland ponies today puts their future in doubt. Their numbers may be less than the critical mass needed for safe reproduction, with population policy left to chance. Beechwood was, essentially, the National Stud of Scotland and without it there is no secure garron bank. The Highland Pony Society cannot possibly fill the realistic role played by the Department of Agriculture throughout the critical latter years of breeding this horse. The pony may have greater market value today than ever before, but this could be the final warning bell, for the value of the animal is supported by a basic, butchering price, determined by a continental trade. The situation does not need to be put to music; in a sense it is already a tune with an old familiar score, mused over by Scott with these words on Caledonia, which could as well apply now to its horse:

> ". . . of all bereft,
> sole friends thy woods and streams are left,
> and thus I love thee better still,
> even in extremity of ill."

If it is no luckier than many other aspects of Caledonian heritage, the days of the garron may indeed be precious.